Lecture Notes in Physics

Edited by H. Araki, Kyoto, J. Ehlers, München, K. Hepp, Zürich
R. Kippenhahn, München, D. Ruelle, Bures-sur-Yvette
H. A. Weidenmüller, Heidelberg, J. Wess, Karlsruhe and J. Zittartz, Köln

Managing Editor: W. Beiglböck

362

Namir E. Kassim   Kurt W. Weiler   (Eds.)

Low Frequency Astrophysics from Space

Proceedings of an International Workshop
Held in Crystal City, Virginia, USA,
on 8 and 9 January 1990

Springer-Verlag
Berlin Heidelberg GmbH

Editors

Namir E. Kassim
Kurt W. Weiler
Center for Advanced Space Sensing, Code 4030
Naval Research Laboratory, Washington, D.C. 20375-5000, USA

ISBN 978-3-662-13774-1   ISBN 978-3-540-47172-1 (eBook)
DOI 10.1007/978-3-540-47172-1

© Springer-Verlag Berlin Heidelberg 1990
Originally published by Springer-Verlag Berlin Heidelberg New York in 1990
Softcover reprint of the hardcover 1st edition 1990

2153/3140-543210 – Printed on acid-free paper

PREFACE

Radio astronomers have been working at low frequencies since the first days of the science. However, the observing limitations and the move to ever shorter wavelengths to achieve higher resolution with fixed dish and array sizes has meant that most areas at low frequencies still remain to be fully exploited with modern techniques and instruments. In particular, the possibilities for pursuing the very lowest frequencies with new ground-based arrays, with interferometry from ground-to-space, with Earth-orbiting instruments, and with arrays on the surface of the Moon promises a rebirth of work in this frequency range. Space or Moon-based efforts are particularly enticing since the effects of the Earth's ionosphere are among the major difficulties confronting low-frequency observations. To explore these possibilities further, a workshop devoted to Low Frequency Astrophysics from Space was held on 8 & 9 January 1990 in Crystal City, Virginia. More than fifty participants from all parts of the U.S. and a number of foreign countries outlined a scientific need for a coordinated ground, earth-orbit, and lunar-surface observatory development program to pursue the many areas of astrophysics which can only be probed at low radio frequencies. These astrophysical areas and instrumental concepts are described in these workshop proceedings.

Unfortunately and quite surprisingly, some astronomers tend to regard low frequencies as an area of astronomy where nothing has ever happened in the past and nothing is likely to happen in the future. Besides being a very narrow mind set which disregards the need of modern astrophysics for observations across the entire electromagnetic spectrum, such a view is quite wrong in all respects. As can be seen from Table 1, the past five decades have seen many of the most exciting discoveries in astronomy made at low frequencies. Now, with new instruments, and, particularly, space and lunar initiatives, the field promises a bright future. The participants at the workshop laid out both a challenging list of astrophysical problems to be solved and a series of exciting instrumental concepts to investigate them. Solar astronomy, planetary science, the thermal interstellar medium, supernova remnants, pulsars, interstellar-plasma refractive and diffractive scattering, cosmic rays, old "fossil" electron populations, quasars, radio galaxies, galactic background and halo studies, coherent emission mechanisms, and possible serendipitous discoveries all promise valuable insights at low frequencies. Also, the simple technology and low cost of low-frequency instruments will allow workers to plan relatively inexpensive and very cost effective programs of ground-based arrays, ground-to-space VLBI, Earth-orbit synthesis telescopes, and large mapping arrays on both the near and far sides of the Moon.

A systematic development plan for such instruments has been assembled by the workshop participants and is presented in a very short overview in Table 2, with more discussion in the proceedings. With the technology now available for opening this last unexplored window on the astrophysics of the Universe, a phased program of ground-based and ground-to-space VLBI observations at frequencies above 10 MHz, and orbiting and lunar-based observatories at lower frequencies, should proceed at once. The 10 and 30 MHz ground-based arrays should be in operation before the next solar minimum and, because of the long lead times for space missions, planning for space-to-ground, Earth-orbit, and lunar arrays should be initiated. Ground-based arrays can also serve as test beds for the hardware and software of space missions. In particular, lunar-array deployment over large areas with remote operation will require extensive ground-based testing. Also, software development of large-field-of-view mapping techniques will be needed for both ground-based and space-based arrays.

In all, the meeting provided a stimulating atmosphere for the exchange of ideas and for the discussion of important scientific possibilities which have too long been ignored by many in the astronomical community. This volume reports these efforts to bring to an end the years of unfortunate neglect of an exciting area of astrophysics.

Center for Advanced Space Sensing   Namir E. Kassim
Naval Research Laboratory   Kurt W. Weiler
Washington DC
May 1990

Table 1: Significant Discoveries in Radio Astronomy at Low Frequencies

Year(s)	Frequency	Discoverer(s)	Short Description
1931-35	20 MHz	Jansky	Discovery of cosmic radio waves: Birth of radio astronomy
1935-40	160 MHz	Reber	First spectrum of galactic background
1940	160 MHz	Henyey, Keenan	Interpretation of Reber's observations requires new nonthermal emission processes
1942	60 MHz	Hey	First detection of solar radio emission
1946	60 MHz	Hey	Meteor stream radiants from radio reflection methods
1946	175 MHz	Ryle et al.	First two element interferometer: Resolved sunspots; defined basic principles of aperture synthesis
1947	200 MHz	Pawsey et al.	Sea interferometer used to resolve sunspots
1946-50	50-150 MHz	Ryle, Hey, Bolton, others	First discoveries of discrete cosmic radio sources (Cyg-A)
1951	50-150 MHz	Ryle, Smith, Baade, Minkowski, Mills, others	Discovery of radio emission from SNRs (Cas-A, Crab Nebula)
1951	50-150 MHz	Ryle, Smith, Baade, Minkowski, Mills, others	Discovery of radio galaxies (Cyg-A, Vir-A, Cen-A, Her-A)
1951	158 MHz	Brown, Hazard	Discovery of radio emission from normal spirals (M31)
1952-53	400 MHz	Mills	First large interferometer: The Mills Cross
1955	85 MHz	Mills	First detection of radio emission from the Magellanic Clouds

Table 1:
(Cont.)

1955	50-250 MHz	Kraus, Mills, Baldwin, others	First all-sky surveys with interferometers
1955	22 MHz 18 MHz	Burke, Franklin Shain	First detection of planetary radio emission (Jupiter)
1955-63	50-200 MHz	Hewish, Vitkovich, Slee, Parker, Högbom	First use of IPS to study the solar corona
1959	400 MHz	Drake and Hvatum, Roberts and Stanley	Evidence for synchrotron emission from Jupiter
1960	50-400 MHz	Blaauw, Gumm, Oort, others	New definition of galactic coordinates based on interpretation of low frequency surveys
1962-63	178 MHz	Bennett	First widely used radio catalogue (3C)
1963	178 MHz	Clark, Hewish	First use of IPS for size limits on compact radio sources
1963	136 MHz 410 MHz	Hazard et al. Schmidt, Sandage Greenstein, others	Discovery of quasars (3C273)
1967	18 MHz	Brown, Carr, Block	First VLBI fringes
1968	85 MHz	Bell, Hewish	Discovery of pulsars -- Nobel Prize

Table 2: Programs for Development of Low Frequency Radio Astronomy

Concept	Institutions[*]	Freq. Range	Short Description
LFVLA (+ VLBA outriggers)	NRAO/NRL	10, 30, 75 MHz	The VLA is currently outfitted with prime focus, 75 MHz dipoles on 4 antennas. Construction is underway to outfit 4 more. Testing and development can be carried out on these with plans being developed for a full 27 element array at 75 MHz. After that, consideration can be given to fully outfitting the VLA and nearby VLBA stations at 10, 30, and 75 MHz.
OLFRAS	NRL/GSFC/ JPL/NMSU	1, 5, 13, 30, 75 MHz	A single satellite in elliptical orbit (~300-3000 km altitude) would work in conjunction with ground based arrays for OVLBI synthesis mapping in the 10-75 MHz range. Interference monitoring could be carried out in the 1-30 MHz range.
LORAE	NMSU/JPL/ NRL/GSFC	15 kHz to 30 MHz	LORAE would be a hitchhiker on the Lunar Observer with one or more low frequency dipoles on one or more orbiters. It would carry out three tasks -- 1. interference monitoring at the lunar distance, 2. occultation mapping of bright sources, and 3. all sky, low resolution surveying.
LFSA	NRL/JPL/ GSFC/NMSU	1, 5, 13, 26 MHz	The LFSA would be an orbiting array in high earth orbit to carry out all sky surveying and high resolution full synthesis mapping of sources in the deka-hectometer wavelength range.
Near Side Lunar Array	GSFC/JPL/ NMSU/NRL/ UTX	150 kHz to 30 MHz	In conjunction with the establishment of the first lunar outpost, it would be possible to place and operate a low frequency array on the lunar near side.
Far Side Lunar Array	GSFC/JPL/ NMSU/NRL/ UTX	150 kHz to 30 MHz	As human presence on the Moon becomes routine and communication links are established from the back side of the Moon to Earth, lunar far side radio astronomy would become possible.

----

[*]Only a few institutions are listed as presently developing initial concepts. Actual programs will certainly involve international cooperative efforts by many more institutions.

CONTENTS

I. PROPOSED LOW FREQUENCY INSTRUMENTS FOR SPACE AND THE MOON

REFLECTIONS ON THE RADIO ASTRONOMY
EXPLORER PROGRAM OF THE 1960S AND 70S

M.L. Kaiser

Goddard Space Flight Center

Greenbelt, MD 20771

The Radio Astronomy Explorer (RAE) program of the late 1960s and early 1970s is, to date, the only totally dedicated radio astronomy mission to have flown. However, only some of the prelaunch goals were achieved due to the unexpectedly high levels of interference from the earth in the form of both naturally occurring and man-made noise. Some important lessons in receiver design were learned which could and should be applied to any future radio astronomy missions.

Mission Goals, Design and Instrumentation

The two Radio Astronomy Explorer (RAE) satellites launched in 1968 and 1973 are often referred to as man's only radio astronomy missions to date. This statement is true only if one adds the word *dedicated*, because low frequency (<10 MHz) radio astronomy receivers have flown on approximately 25 earth-orbiting and interplanetary spacecraft [Kaiser, 1989] . Even though most of these other missions concentrated on planetary and solar science, the radio receiver technology was similar to the RAE receivers providing a considerable body of experience and knowledge in low frequency receiver design.

An overview of the RAE mission results was given by Kaiser [1987] and will only be summarized here. This review will address the successes and failures of the RAE mission with suggestions for future designs. Although the RAE mission will be used as the prime example, many of the statements apply equally well to the other radio astronomy instruments.

The specific goals of the RAE mission were to:

1. determine the cosmic noise spatial and spectral structure
2. measure low frequency solar radio bursts
3. study Jupiter's low frequency noise
4. attempt to detect discrete cosmic noise sources.

These goals were derived from a long history of radio astronomy measurements made by the Goddard Space Flight Center group using sounding rockets and appeared to be achievable with the existing technology and data processing techniques of that era.

In order to accomplish these goals, two nearly identical spacecraft were designed. These two RAE spacecraft each consisted of three antenna systems driving a variety of radio receivers. The antenna systems, shown for the case of RAE-2 in Figure 1, were all dipoles, but two of the dipole systems were configured in the shape of giant Vees, one pointed up and one pointed down (the spacecraft were gravity gradient stabilized). The third dipole system was a more conventional configuration bisecting the two Vees. The purpose of the Vee antennas was to provide

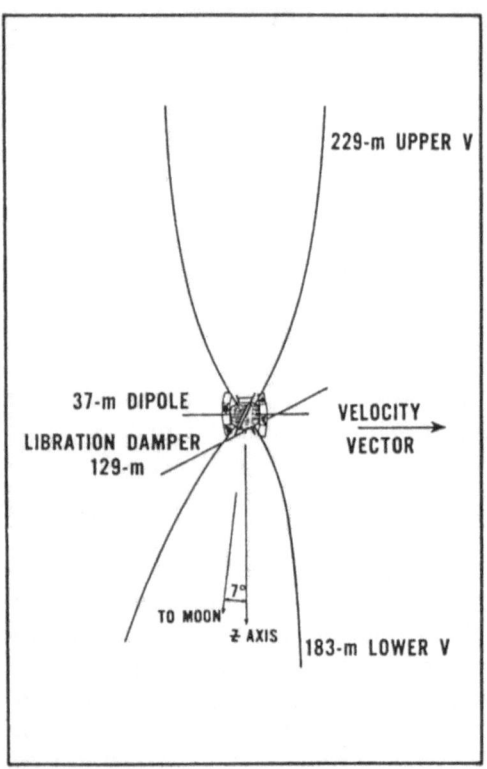

Figure 1. The RAE-2 antenna system showing the bending due to gravity. On RAE-2 the lower Vee was not fully deployed due to a possible mechanical problem.

improved directionality over the normal dipole pattern, and this improvement, although marginal, was achieved as can be seen in Table 1.

Table 1

Radiation characteristics of the RAE Vee antennas

Freq MHz	Beam size deg	Side lobe dB	Front:Back dB
9.18	37 X 61	-2	10
6.55	27 X 55	-4	15
3.93	80 X 63	-5	15
1.31	180 X 120	-12	15
0.87	220 X 160	?	?

The receivers driven by these antenna systems were of two kinds, Ryle-Vonberg radiometers and so-called burst receivers. The important difference between these two types of receivers was in the preamplifier section. The R-V receivers contained narrow-band preamps, one for each observing frequency, whereas, each burst receiver had only one preamp whose bandwidth covered the entire frequency range of the receiver and more. Both types of preamps had operating ranges of from about 1 μV to about 30 mV.

Successes and Failures

The most significant and unexpected discovery of the RAE missions was that the earth itself was a very powerful, even dominating, source of radio emission at low frequencies. RAE-1 was placed in a nearly circular 6000 km earth orbit in 1968, but strong terrestrial emissions, particularly over the night hemisphere and at high latitudes, led to the decision to place RAE-2 in lunar orbit in the hope that the 60 R_E distance would be enough to attenuate the terrestrial noise. Figure 2, from the RAE-2 spacecraft in lunar orbit, dramatically illustrates just how powerful these

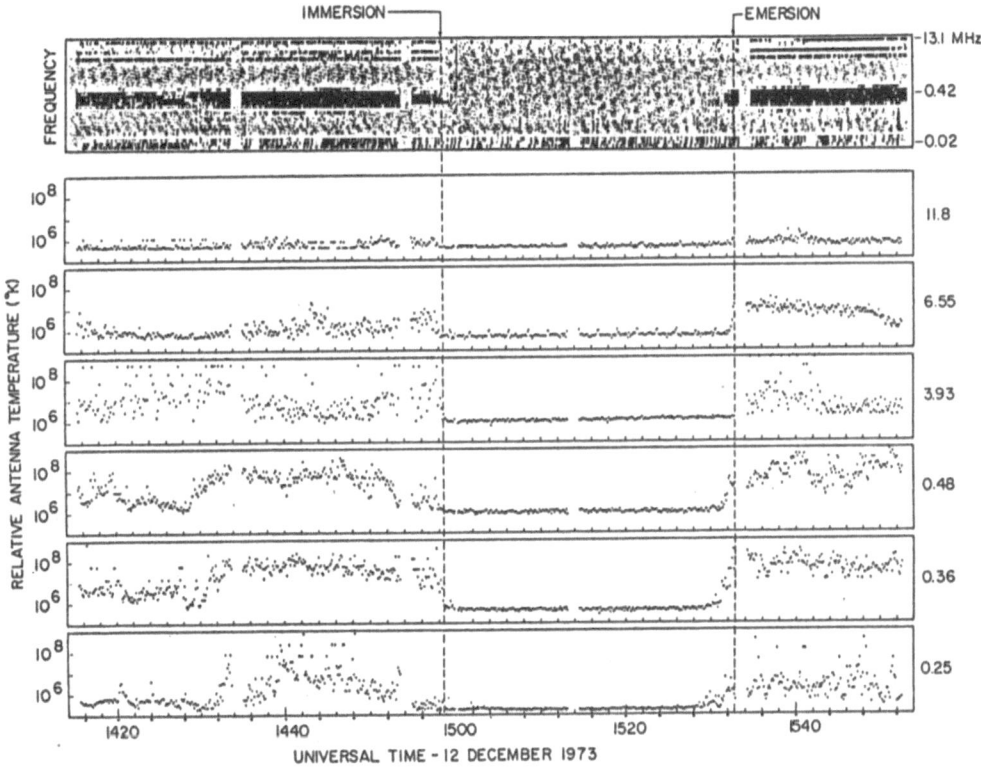

Figure 2. Data from RAE-2 in lunar orbit showing the dramatic disappearance and reappearance of the earth.

terrestrial emissions are. At the top is a dynamic spectrum covering the frequency range of 20 kHz to 13 MHz, with increasing intensity depicted by increasing darkness. The bottom several panels are individual frequency channels displayed as intensity versus time. The indicated immersion and emersion times mark the interval when the earth was out of view from RAE-2 because of blocking by the moon. The extremely intense emission seen at the middle frequencies (150 — 300 kHz) is naturally occurring radio emission from high above the auroral zones now known as AKR or auroral kilometric radiation [see review by Kaiser, 1989]. At the highest frequencies of several MHz, man-made radio noise and spherics from lightning dominate the spectrum. When the moon blocks the view of the earth, all signal levels drop to receiver threshold.

Although the study of AKR itself became an significant part of the RAE mission, the galactic and extragalactic astronomy goals were severely compromised by its presence. The rather large sidelobes (Table 1) and the limited operating range of the preamplifiers made rejection of AKR difficult at best, and nearly impossible most of the time. With sidelobes only 2 to 12 dB below the main beam and front to back ratios of only 10 to 15 dB, virtually any time the earth was visible, regardless of where the Vee antenna was pointed, significant amounts of power entered the receiver. In addition, since AKR is fairly broadbanded ($\Delta f/f$ nearly 100%) and often reaches 10^{-16} W m^{-2}Hz or higher at the spacecraft, the 30 mV maximum allowable input signal in the preamp was easily violated creating non-linear response resulting in apparent signal enhancements at frequencies well separated from the AKR band.

Compared to the original list of goals, one would have to say that the RAE mission was only partially successful, meeting goals 2 and 3, but having essentially failed goals number 1 and 4. However, nearly 25% of the approximately 60 published scientific papers associated with RAE concerned the earth as a radio source, a subject not even on the original list of goals. Thus, from an overall scientific standpoint the RAE mission was very successful.

Some Suggestions for Future Radio Astronomy Receivers

Scientifically, RAE goals 1 and 4 still are very appealing. The low frequency radio spectrum, from the standpoint of galactic and extragalactic astronomy, represents one of the last unexplored regions of the electromagnetic spectrum. This region, below 10 MHz, probably contains as many new phenomena as any other spectral region. The problem is, how do we explore it? This spectral region poses different problems from virtually any other spectral region. Usually, a region remains unexplored because adequate sensitivity is not available. In the low frequency radio astronomy band, receiver sensitivity is not a problem – all signals are very strong compared to typical radio receiver sensitivities. The problem is somewhat the opposite, i.e., too much signal, mostly from unwanted sources. Clearly, as can be seen from figure 2, the far side of the moon is a very attractive place to conduct low frequency observations. Complete avoidance of terrestrial interference is the ultimate solution to the low frequency radio astronomy problem.

This does not mean, however, that progress on galactic and extragalactic astronomy is impossible before a far side lunar observatory is built. Some of the lessons learned from RAE and other radio astronomy instruments can be put into practice. Perhaps most important is in the area of preamplifier design. Either individual narrowband preamps should be used, one for each frequency band, or a much larger dynamic range broadband preamp , good to several volts, should be employed. This will allow strong signals to be within the observing band without creating false signals.

The use of the moon as a large occulting disk is still a viable option. With the RAE-2 spacecraft, the combination of low dynamic range preamp and very short integration time (tens of milliseconds) made detection of discrete sources other than the sun and planets (earth included) difficult, but not impossible. Some unpublished work from RAE-2 by this author showed that a few of the more intense discrete radio sources (e.g. Fornax-A, Cygnus-A) were detectable via lunar occultation in spite of the severe problems. A redesigned receiver with a large dynamic range preamp and integration time constant more appropriate to occultation time constants (i.e. a sizeable fraction of the first Fresnel zone – several seconds instead of milliseconds) could detect, in principle, several hundred discrete radio sources at 1 MHz.

I believe it is important that we not "give up" on low frequency astronomy. I feel we can solve the technical problems and that the potential scientific rewards are well worth the rather modest investment required.

References

Kaiser, M.L.: 1987, Radio Astronomy from Space, NRAO Workshop No. 18.

Kaiser, M.L.: 1989, in Plasma Waves and Instabilities at Comets and in Magnetospheres, AGU Geophysical Monograph No. 53.

LOW FREQUENCY ASTROPHYSICS WITH A SPACE ARRAY

Kurt W. Weiler

Center for Advanced Space Sensing

Naval Research Laboratory, Washington, DC 20375-5000

ABSTRACT

There is a strong astrophysical need for surveying the entire sky and imaging individual sources at frequencies between ~1 and ~30 MHz, a frequency range over which the Earth's ionosphere transmits poorly or not at all. High resolution, high sensitivity observations would open a new window in the electromagnetic spectrum for astronomical investigations and bring astronomy to the fundamental physical limit below which the Milky Way becomes optically thick over relatively short path lengths due to diffuse free-free absorption. We discuss here the scientific questions and initial instrumental concepts for a space mission to carry out such work by combining low cost and low complexity with high productivity of new scientific results.

TEXT

1 Introduction

The opening of a new spectral window for astronomical investigations has always resulted in major discoveries, significant insights into astrophysical processes, and an enrichment of our understanding of the universe. The present discussion is directed towards investigation of the possibility of an interferometric, space-based array for imaging the entire sky at frequencies below 30 MHz, a range which is totally inaccessible or extremely difficult to observe from the ground due to ionospheric absorption and scattering. Because of this large gap in our knowledge, even though many worthwhile projects can already be defined for a new instrument, the likelihood of discovering new processes and objects is great. An improvement in sensitivity to the few Jansky level and in resolution to the sub-minute-of-arc level would be as much of an advance for the field as was the Einstein satellite for x-ray astronomy or the IRAS for infrared astronomy. Also, observing at frequencies as low as ~1 MHz extends astronomy to the lowest practicable physical limit for studying electromagnetic radiation from within our Galaxy. At still lower frequencies the diffuse interstellar ionized hydrogen gives a very high optical depth due to free-free absorption over relatively short path lengths (Alexander, et al. 1969).

2 A Low Frequency Space Array (LFSA)

In order to develop a set of scientific goals one must first estimate the likely instrumental parameters and capabilities. An initial instrumental concept might be that of a single spacecraft bus placing four to eight free-flying satellites (array elements of a LFSA) into very similar circular orbits of large radius (>20,000 km) at high inclination. The bus itself would play no role other than providing final orbit injection and would be kept as simple and inexpensive as possible. The array elements would all be identical, unstabilized spacecraft, each with three mutually orthogonal, broad beam, wide bandwidth dipole antennas. The receivers would be near 1.5, 4.4, 13.4, and 25.6 MHz with individual bandwidths of ~50 kHz. The signals received by all dipoles, after suitable digitization and the addition of monitor and control information, would be transmitted in real time to the ground from each array element for recording and later archiving, correction, correlation, mapping, and analysis. Clock stability aboard each array element would be such that full array coherence is maintained at all times.

The orbits for the array elements would be chosen such that differential dynamical forces cause the array to expand, without active adjustment, from an initial element separation of <1 km to a final separation of >300 km (resolution ~10' at 1.5 MHz to resolution ~10" at 25.6 MHz) while the orbital revolution and precession sweep out a survey of the sky. After a period of ~1 year of coherent integration with varying satellite-to-satellite baseline lengths and orientations, a complete synthesis of the sky with high resolution and sensitivity at all 4 frequencies would be available. Since each array element would need to have minor orbit adjustment capability (micro-thrusters), at the end of a surveying interval orbital adjustment would be made such that the array recompresses for repeating, at least once and hopefully twice, the survey to obtain error and consistency checks.

An estimated capability of an LFSA with one possible antenna design is given in Tables 1 and 2.

Table 1: Estimated Capability

Freq. (MHz)	T_s (K)	t (sec)	A_e (m^2)	σ (Jy)	N_{det}	Appx. Resol.[a]
1.5	3×10^7	7.5×10^6	2400	40	1300	~0.°1
4.4	5×10^6	3.8×10^6	1000	9	5000	~1'
13.4	3×10^5	1.5×10^6	400	3	6000	~0.'3
25.6	3×10^4	$\sim 10^6$	175	1.5	10000	~10"

[a]The maximum resolution is generally limited by the interstellar scattering (see Table 2).
T_s =Effective System Temperature (determined by Galactic background)
t =Integration Time = (0.5 data loss factor) x 1 year/directivity
A_e =Total Effective Array Aperture
σ =Rms Error (assuming 50 kHz bandwidth per channel, and a sensitivity constant of 2)
N_{det} =Number of Detectable Sources (extrapolated from the Clark Lake survey; Viner and Erickson, 1975)

Table 2: Estimated Scattering Diameters

Frequency (MHz)	Scattering Diameter (arcsec)	Max. Useful Baselines (km)
1.5	300 - 1800	20 - 130
4.4	50 - 150	100 - 300
13.4	6 - 20	300 - 1000
25.6	1 - 3	1000 - 3000

3 Scientific Programs

An astrophysical observatory operating at very low frequencies would be a powerful tool for investigating a broad range of phenomena. Here we discuss some of the most important questions to be addressed.

3.1 Galactic Nonthermal Background

Study of the distributed nonthermal background emission of the Milky Way is a rewarding task in all parts of the electromagnetic spectrum; different frequencies emphasize different physical processes. For example, the COS-B survey of γ-ray emission is sensitive to the interaction of cosmic rays with the ambient interstellar gas, the well known Palomar Sky Survey emphasizes stars and ionized hydrogen (HII) regions, and the IRAS survey enhances visibility of the relatively cold interstellar dust. Radio frequency studies (see, e.g., Haslam, et al., 1982 at 408 MHz; Wielebinski, Smith, and Cardenas, 1968 at 151 MHz; Milogradov-Turin and Smith, 1973 at 38 MHz; Kassim, 1988 at 30.9 MHz; and Jones and Finlay, 1974 at 29.9 MHz) are known to be most sensitive to the relativistic electron component of cosmic rays and the distribution of interstellar magnetic fields.

By performing surveys at the low frequencies and high resolutions which would be possible with an LFSA, we would not only be able to define galactic background radiation spectral indices below the known 300 MHz break to high precision but would also be able to resolve and study the distribution of different components. For example, it should be possible to test the theories on whether the loop-like features seen in the background are old remnants from nearby supernovae (Berkhuijsen, 1971) or loops of magnetic field and particles leaking out of the galactic plane (Parker, 1965) by measuring their spectral index values and distributions. Also, at low frequencies the non-thermal halo of the Milky Way will be prominent and it should be possible to measure its extent, relativistic particle density, and magnetic field strength for a better determination of the confinement mechanisms for cosmic ray e$^-$ and p$^+$.

3.2 Galactic Diffuse Free-Free Absorption

By observing a large number of extragalactic radio sources and determining their low frequency spectra as a function of galactic latitude and longitude, it should be possible to measure the changes due to absorption by the diffuse, interstellar gas in the Milky Way and thereby investigate its distribution. Then, models for a "warm" ($T_e \sim 10^4$ K) disk of ionized hydrogen imbedded in a "hot" ($T_e \sim 10^6$ K) halo could be tested and a global picture of the free-free absorption obtained for comparison with existing higher frequency pulsar dispersion and Faraday polarization rotation measurements. These results would also be supplementary to the COS-B γ-ray measurements which are related to the local cosmic ray energy and interstellar gas density.

A complementary technique to studying the diffuse free-free absorption could be to use nearby HII regions, which are optically thick at these low frequencies, to completely block the nonthermal background radiation from more distant parts of the Galaxy. Then, any emissivity observed could be attributed solely to the synchrotron radiation arising between the observer and the HII region, yielding a measurement of local values of the cosmic ray e$^-$ and magnetic field components.

3.3 Interstellar Scattering and Refraction

It is generally accepted that small scale ($\sim 10^9$ cm) fluctuations in electron density in the interstellar medium can diffractively scatter radio waves from a background source (see, e.g., Dennison, et al., 1984). Less clear, yet of considerable importance, is the ability of somewhat larger irregularities ($\sim 10^{13}$ cm) to refractively focus and defocus radio waves. Such refractive scintillation has been proposed as the origin of some low-frequency variability observed in compact extragalactic sources (Shapirovskaya, 1978; Rickett, Coles, and Bourgois, 1984). Because the scattering and refraction angles scale roughly as $\lambda^{2.2}$, these effects will be quite large at low frequencies, causing measurable angular broadening of many of the discrete sources in the all-sky surveys at 1.5, 4.4, and 13.1 MHz. With many broadening measurements spanning the full range of galactic latitudes and longitudes, it will be possible to map the distribution of scattering material in the Milky Way.

3.4 Extragalactic Sources

Very little is known about the properties of individual radio sources at low frequencies. However, the expected capability of an LFSA is such that thousands (see Table 1) of discrete sources could be detected and the brighter ones studied for such properties as integrated spectrum, surface brightness and spectral index distribution, and source counts ("log N - log S"). This is especially important since the relativistic electrons which the LFSA would detect are, in general, far older than those normally studied by radio astronomy. Thus, "fossil," steep spectrum sources which are not observable at higher frequencies may become available for study with an impact on theories for the evolution and lifetimes of radio sources and of the universe.

3.4.1 Spectra

Radio sources are generally characterized by a spectral index α which describes the change of flux density S with frequency as a power of the observing frequency ν ($S \propto \nu^{+\alpha}$). However, it has long been known that the spectral index is a function of frequency $\alpha(\nu)$ and that the measurement of this frequency dependence is important for understanding the physics of the emission and absorption processes and the physical environment in the sources. Determination of $\alpha(\nu)$ requires measurements of source flux densities at all frequencies, and, in particular, at low frequencies where a number of absorptive and emissive processes become prominent. At present, very little is known about source spectra at frequencies as low as 20 MHz and practically nothing has been measured for $\nu < 10$ MHz. It is thus an important parameter of an LFSA that it would have multifrequency capability to provide this information throughout this poorly explored range and to investigate the many possible emission and absorption processes involved.

3.4.1.1 Steep Spectrum (Emission Processes)

Enhanced emission from sources at the lowest frequencies (which we will collectively call "steep spectrum") and thereby enhanced numbers of individual sources available for study can be anticipated due to a number of phenomena. In addition to the possibility that old "fossil" electron components would become visible, there is known to exist a class of sources with spectral indices $\alpha < -1$ which are only detectable at low frequencies (see, e.g., Baldwin and Scott, 1973). These display no peak in their spectra and apparently increase monotonically in flux density even at the lowest measured frequencies.

Some low frequency sources are quite large (~1 Mpc) and are associated with clusters of galaxies. Erickson, Matthews, and Viner (1978) have shown that there is, in fact, a strong correlation between these steep spectrum clusters and x-ray emission from the intra-cluster gas. However, the physical connection remains unclear. A possibly related question is the process by which central radio galaxies in some clusters "leak" their relativistic electrons into the intra-cluster medium to produce a cluster "halo" of megaparsec dimensions such as that seen in the Coma Cluster. Hanisch and Erickson (1980) have shown that the apparent rate of relativistic particle propagation observed in clusters which are bright at low frequencies is several orders of magnitude greater than the Alfven speed and even exceeds the ion sound speed. These high propagation rates seem to rule out a diffusion process and are consistent with a constant velocity streaming of the particles. However, this is surprising since the intra-cluster medium is expected to be quite inhomogeneous and stirred by galaxy motions.

3.4.1.2 Peaked Spectrum (Absorption Processes)

Many radio sources display a spectrum which peaks in flux density at some intermediate frequency and declines both above and below that. For extended sources the frequency of peak flux density is usually quite low and in some cases this "turnover" is presumed to occur at frequencies below present observational limits. Since for all of these sources the emission process is assumed to be synchrotron radiation with a constant power law index, a decline at low frequencies is usually attributed to one of three absorption processes: external ionized hydrogen free-free absorption, internal free-free absorption, and synchrotron self absorption. Since these processes have different and predictable frequency behaviors, accurate measurement of low frequency spectra could determine which is prominent (see, e.g., Weiler 1987 and Weiler et al. 1988). These three processes, at least for simple models, have distinctive signatures below the turnover frequency ν_r. This is illustrated in Figure 1.

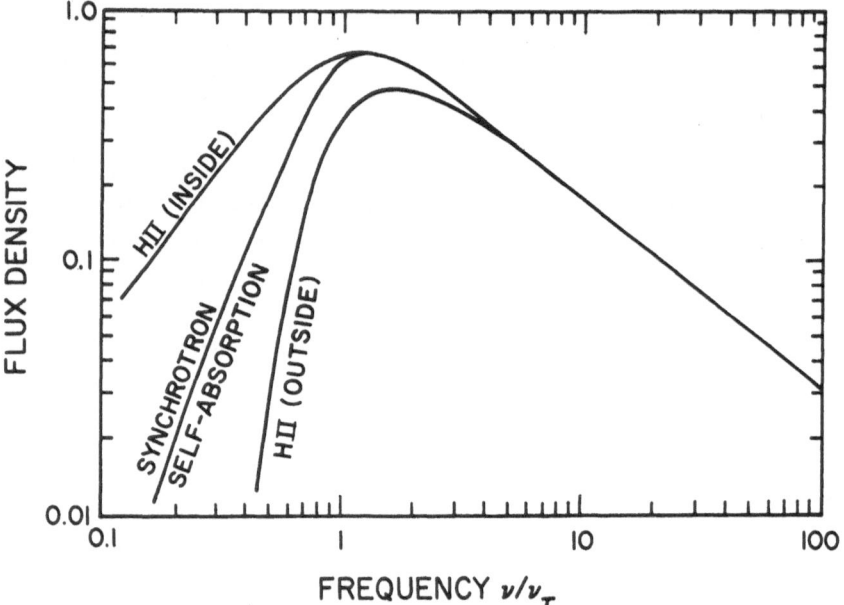

Figure 1: Schematic representation of the frequency dependence of three possible absorption mechanisms below spectrum turnover at $\sim\nu_r$. Scales are in relative units.

3.4.1.3 Bent Spectrum (Source Evolution)

A number of loss processes can affect the energy of relativistic electrons. These alter the observed spectrum in different ways. For example (Longair, 1981), if γ is the power law index of the cosmic ray electron energy spectrum $N(E) \propto E^{-\gamma}$ [N.B.: It can be shown that this cosmic ray energy index γ is related to the radio spectral index α by $\gamma = (1-2\alpha)$.] and if ionization losses dominate, $N(E) \propto E^{-(\gamma-1)}$, i.e., the observed spectral index α is flatter by 0.5. If bremsstrahlung or adiabatic losses dominate, $N(E) \propto E^{-\gamma}$, i.e., the spectral index α is unchanged. If inverse Compton or synchrotron losses dominate, $N(E) \propto E^{-(\gamma+1)}$, i.e., the index α is steeper by 0.5. Thus, if an estimate can be made of the dominant energy loss processes and magnetic field strengths, detection of breaks in the spectrum permits the determination of the approximate age of the relativistic electrons being observed.

3.4.1.4 Cutoff Spectrum (Injection Processes)

Another possible source of spectral turnovers at low frequencies is the requirement that the acceleration mechanisms for relativistic electrons become inefficient at some low energy. Since the number of relativistic electrons increases very rapidly with decreasing energy ($\gamma \sim 2.5$), at some point this process must cease to avoid an infinite energy content for radio galaxies. Searches have been made (Erickson and Cronyn, 1965; Hamilton and Haynes, 1968; Roger, Bridle, and, Costain, 1973) but such a cutoff has never been clearly identified. To investigate this further, determination of spectra down to the lowest frequencies is required.

3.4.2 Brightness Distributions

Numerous sources detectable with an LFSA would be resolved. At these 3 to 4 order of magnitude lower frequencies diffuse synchrotron components have very high brightness temperatures and the detailed mapping of supernova remnants, normal galaxy disks, normal galaxy halos, radio galaxies and quasars, head-tail sources, component bridges, and distributed emission from clusters of galaxies to deeper levels and greater extensions would be possible. An LFSA would also be able to map spectral index distributions across extended objects to high accuracy permitting searches for evolutionary effects in component motions and/or electron distributions such as Perley and Erickson (1979) have been able to do for giant radio galaxies and Winter, et al. (1980) have been able to do for the bridges in Cygnus A. Such information would help to identify relativistic electron injection, acceleration, diffusion, and evolution processes which are still only poorly understood.

3.5 Pulsars

The spectra of most pulsars turn over in the 100 to 500 MHz range, but a few, interesting, fast pulsars have spectra which are very steep and flux densities which continue to increase down to the lowest observed frequency of 10 MHz. Two examples are the Crab pulsar (PSR0531+219; Bobeiko, et al., 1979) and the millisecond pulsar (PSR1937+214; Erickson and Mahoney, 1985). These pulsars are among the strongest sources in the sky at 10 MHz so that LFSA observations might be able to discover other, similar objects. Also, to avoid radiating infinite power, the spectra of these pulsars must turn over at some frequency <10 MHz and measurement of this turnover frequency could provide information on the spatial structure of the coherently radiating electrons in the pulsar's magnetosphere.

3.6 Coherent Emission

A very exciting possibility at low frequencies is the detection of coherent radiation processes. There are valid physical reasons to anticipate that the smaller distance between individual radiating electrons measured in terms of the electromagnetic wavelength is likely to amplify collective radiative modes. In such a plasma, the ratio of stimulated emission to spontaneous emission can be very high, varying as ν^{-3}. If an inverted energy level population can be established and is sufficiently long lived, there are numerous collective modes which can be excited by instabilities in the magnetoactive plasma. In fact, the occurrence of coherent emission at low frequencies appears to be the rule rather than the exception for solar system objects such as the Sun, the major planets, and the magnetosphere of the Earth. Since objects such as the Crab Nebula, Seyfert galaxies, and quasars typically have densities of $\sim 10^4$ to 10^5 cm^{-3} and magnetic fields <1 Gauss, we might anticipate an analogous situation leading to coherent plasma phenomena in the 1 to 3 MHz range.

3.7 Solar System Observations

Although not our primary area of interest, any deep survey of discrete sources with an LFSA would of necessity reveal Jupiter and the Sun as the brightest radio emitters in the sky at all frequencies under consideration. This would, therefore, present the opportunity of studying them and other solar system bodies with high resolution and provide considerable and heretofore unobtainable information regarding the structure of their emission regions, their plasma properties, and their emission mechanisms.

3.7.1 Sun

In the 1 to 10 MHz frequency range, the effective surface of the Sun ranges from 3 to 30 R$_\odot$. From the active corona, intense emission is to be expected at these low frequencies. For the long lived emission centers, an LFSA would allow mapping of the propagation of electron streams through the corona (Type III emission) and propagation of coronal shock waves (Type II emission).

3.7.2 Jupiter

Extensive Earth-based and spacecraft-based observations have revealed the rich phenomenology of Jupiter's nonthermal emission (Carr, Desch, and Alexander, 1983). However, owing to disturbance by the ionosphere, no direct information exists on its location within the jovian magnetosphere. In a major review of the field, Goldstein and Goertz (1983) stress the importance of position determinations for theoretical understanding of the emission processes.

4 Preliminary Spacecraft Estimate

4.1 General Array Properties

An LFSA is envisioned as an array of four to eight independent, unstabilized array elements orbiting together to form a coherent interferometer. At least four array elements are needed to provide a minimum instantaneous mapping capability for possibly time variable phenomena and, more importantly, for permitting the use of such data correction and calibration techniques as phase and amplitude closure. More elements would provide enhanced phase restoration capability, increased sensitivity, better snapshot capability, and redundancy in case of the failure of one or more elements. All array elements would be identical and kept as simple as possible to give maximum reliability and cost effectiveness. No onboard recording is envisioned and the entire received bandwidths, after digitization, would be transmitted to the ground in real time and recorded on VLBI tapes for archiving and for later correction, cross correlation, and mapping. Each element would carry a clock which is sufficiently stable between updates to maintain the entire array in phase coherence at all times.

The antennas on each array element would provide broad beamwidth and wide frequency bandwidth with good efficiency at all of the planned operating frequencies and with effective collecting areas large enough to permit both distributed Galactic background radiation mapping and individual source imaging. The antennas must also have stable beam shape and gain characteristics. The physical structure of the antennas should be as compact and rigid as possible and easily deployed from the spacecraft.

Because the system temperature would always be dominated by the Galactic background radiation (see Table 1), the receiver temperature is relatively unimportant. State-of-the-art components would be used, but no cooling or special developments are needed.

The array element orbits would be chosen such that they precess as rapidly as possible and, due to differential dynamical forces, slowly expand from an initial separation of elements of <1 km to maximum baselines of >300 km over a period of ~1 year without active control. This would provide the dense coverage of baseline lengths and orientations necessary for high quality synthesis mapping of the whole sky. The inclination of the orbits would be chosen high enough that full sky coverage is produced by the orbital revolution and precession with time. The radius of the orbits would be large enough to avoid the extreme radiation environment of the Van Allen Belts and to minimize the residual magnetospheric effects on the incident radio radiation. Some minor orbit adjustment capability would be provided for correcting non-dynamical effects and for re-initializing the array after maximum baselines are reached.

Figure 2: Artists conception of an LFSA with six array elements.

4.2 Individual Antennas

Although travelling wave "V" antennas were used on the Radio Astronomy Explorer (RAE) satellites (Weber, Alexander, and Stone, 1971; Duff, 1964; Iizuka and King, 1965), their long (229 m) and "floppy" arms and resulting unstable beam shapes and gains, argue against their use for an LFSA. They are also likely to be too expensive. We plan to investigate the use of 3 mutually perpendicular dipoles (see Mahoney et al. 1987), each ~20 m in length (tip-to-tip) which are easier to deploy, less expensive, and give good beam stability. An artists conception of such an array is shown in Figure 2. No active pointing of the antennas is planned, and the beams would, in fact, scan the sky as the unstabilized array elements slowly tumble.

4.3 Sensitivity

With the full sky coverage of its antennas, revolution of the array around its orbit, precession of the orbital plane, and slow array expansion, an LFSA would image the entire sky to high resolution over the course of ~1 year. Because it would be maintained in phase coherence, this would also provide a very long effective integration time and would give very good sensitivity with the large number of detectable sources shown in Table 1.

4.4 Receivers

The receiver noise temperatures would be determined entirely by the Galactic background radiation. Therefore, the main requirements are that the receivers be simple, reliable, and very stable. The receiver center frequencies have been chosen to cover the range from the lowest possible above the diffuse interstellar ionized hydrogen cutoff (~1.5 MHz) to the point where ionospheric limitations become much less severe for ground based observations (25.6 MHz), with steps no larger than factors of 2 to 3 between bands (4.4 and 13.4 MHz). Additionally, near 25.6 and 13.4 MHz, there exist protected radio astronomy bands which would be utilized to reduce problems with man-made interference.

Receiver bandwidths of 50 kHz are considered as a compromise between the susceptibility to interference of wide bandwidths and the low sensitivity of narrow bandwidths. Remote tunability of the center frequency of each receiving band of ~10% is desirable in order to move away from possible interfering signals and each band would likely have to be split into smaller subbands, again for interference suppression.

4.5 Orbits

Orbital parameters would be chosen for optimum stability of both array form and rate of expansion and precession and to provide good quality imaging of the sky both instantaneously and over the mission lifetime. An orbital shape which is circular to a high degree (E < 0.001) is desirable for establishing and maintaining the array shape, stability, and rate of drift expansion. The orbital radii (A) should be large enough to minimize magnetospheric plasma effects on the array and to avoid the severe radiation background of the Van Allen Belts (>20,000 km). The orbit inclination should be low enough to provide reasonably rapid nodal precession but high enough to provide full Sun exposure for long periods to minimize the loss of power and sharp thermal effects from Sun-shadow transitions. An ability to make accurate minor adjustments to the orbital parameters of individual array elements for establishing, maintaining, and recompressing the array ($\Delta A \sim 1$ m; $\Delta E \sim 0.001$; $\Delta I \sim 0.001$ deg) is necessary.

Although the number of possible orbital parameters is large and optimization of the orbits for multiple elements needs to be studied, computer simulations such as those reported in Weiler at al. (1988) show that configurations exist which remain stable, have excellent imaging capability, and expand in a systematic manner. The Fourier (u,v) plane coverage and synthesized beam shape for one 4 element, 10,000 km radius array is shown in Figure 3.

4.6 Space-Ground Link

In order to keep the costs of the program low, it is envisioned that observations would be taken only when the array is in view of a ground station and that no onboard recording capability would be provided. While this is somewhat limiting, it considerably reduces the cost and complexity of the array elements and increases their reliability. By making the ground stations small, cheap, and automated, it should be possible to provide a large number to give good coverage of the surface of the Earth and continuous or almost continuous observing capability.

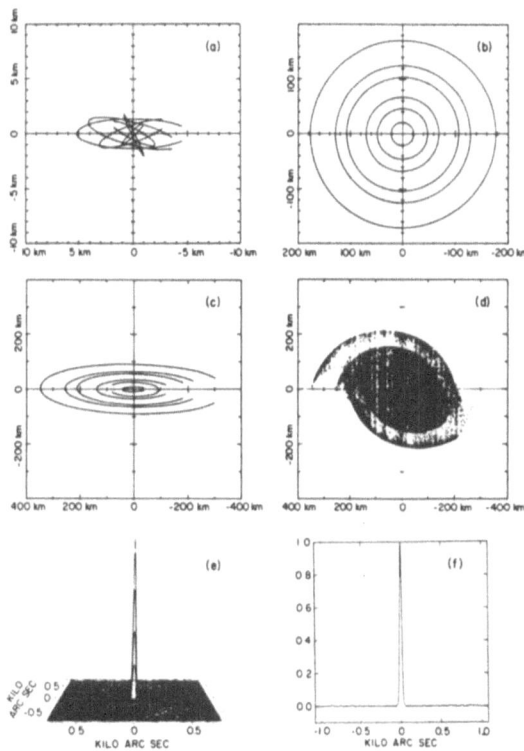

Figure 3: Fourier (aperture) plane coverage by a four element LFSA with a 10,000 km orbital radius for a source at R.A. = 6^h, Dec. = 45° during one orbit (~$3.^h5$) on the day of launch (Day 1; Figure 3a), ~6 months later (Day 180; Figure 3b), and ~1 year later (Day 360; Figure 3c). The total aperture plane coverage for all orbits during 1 year by the slowly expanding array is given in Figure 3d and its Fourier transform (i.e., the resulting point spread function) is shown in Figure 3e. A cross cut through the point spread function, to better illustrate the resulting sidelobe levels, is given in Figure 3f. (N.B.: The scales are different for different figures.)

5 References

Alexander, J.K., Brown, L.W., Clark, T.A., Stone, R.G., and Weber, R.R. 1969, Ap.J. Lett. 157, L163.

Baldwin, J.E., and Scott, P.F. 1973, M.N.R.A.S. 165, 259.

Berkhuijsen, E.M. 1971, A&A 14, 359.

Bobeiko, et al. 1979, Ap. and Sp. Sci. 66, 211.

Carr, T.D., Desch, M.D., and Alexander, J.K.,1983, in Physics of the Jovian Magnetosphere, ed., A.J. Dessler, (Cambridge Univ. Press, Cambridge), p. 226.

Dennison, B., Thomas, M., Booth, R.S., Brown, R.L., Broderick, J.J., and Condon, J.J. 1984, A&A 135, 199.

Duff, B.M. 1964, in The Resistively-Loaded V-Antenna, NSG-579 Sci. Report #3, (Harvard Univ., Cambridge).

Erickson, W.C. and Cronyn, W.M. 1965, Ap.J. 142, 1156.

Erickson, W.C., and Mahoney, M.J. 1985, Ap.J. Lett. 299, L29.

Erickson, W.C., Matthews, T.A., and Viner, M.R. 1978, Ap.J. 222, 761.

Goldstein, M.L., and Goertz, C.K. 1983, in Physics of the Jovian Magnetosphere, ed., A.J. Dessler, (Cambridge Univ. Press, Cambridge), p. 317.

Hamilton, P.A., and Haynes, R.F. 1968, Aust. J. Phys. 21, 895.

Hanisch, R.J., and Erickson, W.C. 1980, A.J. 85, 183.

Haslam, C.G.T., Salter, C.J., Stoffel, H., and Wilson, W.E. 1982, A&A Suppl. 47 1.

Iizuka, K., and King, R.W.B. 1965, in The Travelling-Wave V-Antenna, NSG-579 Sci. Report #4, (Harvard Univ., Cambridge).

Jones, B.B., and Finlay, E.A. 1974, Aust. J. Phys. 27, 687.

Kassim, N.E. 1988, Ap.J. Suppl. 69, 715.

Longair, M.S. 1981, High Energy Astrophysics, (Cambridge Univ. Press, Cambridge).

Mahoney, M.J., Jones, D.L., Kuiper, T.B.H., and Preston, R.A. 1987, Low Frequency VLBI in Space Using "Gas-Can" Satellites, JPL Publ. No. 87-36.

Milogradov-Turin, J., and Smith, F.G. 1973, M.N.R.A.S. 161, 269.

Parker, E.N. 1965, Ap.J. 142, 584.

Perley, R.A., and Erickson, W.C. 1979, Ap.J. Suppl. 41, 131.

Rickett, B.J., Coles, W.A., and Bourgois, G. 1984, A&A 134, 390.

Roger, R.S., Bridle, A.H., and Costain, C.H. 1973, A.J. 78, 1030.

Shapirovskaya, N. Ya. 1978, Soviet Astron. 22, 544.

Viner, M.R., and Erickson, W.C. 1975, A.J. 80, 931.

Weber, R.R., Alexander, J.K., and Stone, R.G. 1971, Radio Science 6, 1085.

Weiler, K.W. 1987, Radio Astronomy From Space, NRAO Green Bank Workshop No. 18.

Weiler, K.W., Dennison, B.K., Johnston, K.J., Simon, R.S., Erickson, W.C., Kaiser, M.L., Cane, H.V., Desch, M.D., and Hammarstrom, L.M. 1988, A&A 195, 372.

Wielebinski, R., Smith, D.H., and Cardenas, X.G. 1968, Aust. J. Phys. 21, 185.

Winter, A.J.B., Wilson, D.M.A., Warner, P.J., Waldram, E.M., Routledge, D., Nicol, A.T., Boysen, R.C., Bly, D.W.J., and Baldwin, J.E. 1980, M.N.R.A.S. 192, 931.

THE LUNAR OBSERVER RADIO ASTRONOMY EXPERIMENT (LORAE)

Jack O. Burns
Department of Astronomy
New Mexico State University
Las Cruces, NM 88003

ABSTRACT: We propose to place a simple low frequency dipole antenna on board the Lunar Observer (LO) satellite. LO will orbit the Moon in the mid-1990's, mapping the surface at high resolution and gathering new geophysical data. In its modest concept, LORAE will collect crucial data on the radio interference environment while on the near-side (to aid in planning future arrays) and will monitor bursts of emission from the Sun and the Jovian planets. LORAE will also be capable of lunar occultation studies of >100 of the brightest sources, gathering arcminute resolution one-D data on sizes and measuring source fluxes. A low resolution all-sky map below 10 MHz, when combined with data from the Gamma-ray Observatory, will uniquely determine the density of galactic cosmic ray electrons and the strength of the Galaxy's magnetic field. LORAE also will be able to measure the density of the Moon's ionosphere -- presently poorly known but important for future lunar-based low frequency arrays. This technically simple, low cost, light weight experiment could provide a significant first step in space-based low frequency astronomy.

INTRODUCTION

On July 20, 1989, the twentieth anniversary of the Apollo 11 landing on the Moon, President Bush committed the U.S. to the establishment of a permanently manned lunar outpost. With this commitment came the recognition that astronomical observatories on the lunar surface (e.g., Burns and Mendell, 1988) could become a reality by the early 21st century. Among the proposed observatories is a lunar far-side very low frequency array (Douglas and Smith, 1985; Burns et al., 1990; Basart and Burns, 1990). It is generally agreed that the radio-quiet of the far-side will provide the best opportunity for sensitive surveys of celestial sources below 10 MHz. Since it is likely that the lunar near-side will be the first location for a manned outpost, a far-side array may be not be possible until a later phase of development -- sometime in the 2010's. However, we may not have to wait this long to gain access to the lunar far-side. Serious planning is underway for a Lunar Observer satellite (a precursor to the lunar outpost), to be launched in the mid-1990's. It will orbit the Moon gathering new geophysical data and mapping the surface at high resolution.

In this paper, we propose a low frequency radio astronomy experiment for the Lunar Observer. LORAE offers an opportunity to expeditiously perform important low frequency observations in the radio-quiet environment of the Moon at modest cost and to accurately measure the interference environment toward Earth. This is an important first mission in an evolutionary low frequency observatories scheme that will lead to a lunar far-side array.

ADVANTAGES OF THE MOON

RAE-2, launched in 1972, demonstrated that the lunar far-side is the lowest radio noise environment in the Earth-Moon system (e.g., Kaiser, 1990). This Explorer-class satellite was placed in lunar orbit. The dynamic spectrum shown in Fig. 1 graphically illustrates the dramatic drop in radio interference experienced by RAE-2 when it orbited the lunar far-side. At night on the far-side, the background from 250 kHz to 12 MHz is that expected from Galactic low frequency radiation. In contrast on the near-side, interference from the Earth totally dominates. Above 1 MHz, Earth-based interference arises from ionospheric breakthrough of man-made radio transmissions (especially above 10 MHz) and terrestrial atmospherics. Below about 1 MHz, interference is produced by the auroral kilometric radiation (AKR) first discovered by RAE (e.g., Kaiser, 1990).

Even on the near-side, the Moon has advantages over high Earth orbit (HEO, >10,000 km) because of the greater distance from the source of interference. On the near-side, interference is reduced by -30 to -35 dB in comparison to HEO (although it is still quite significant).

The final advantage is that the recent lunar initiative may provide us with access to the Moon within this decade. The Lunar

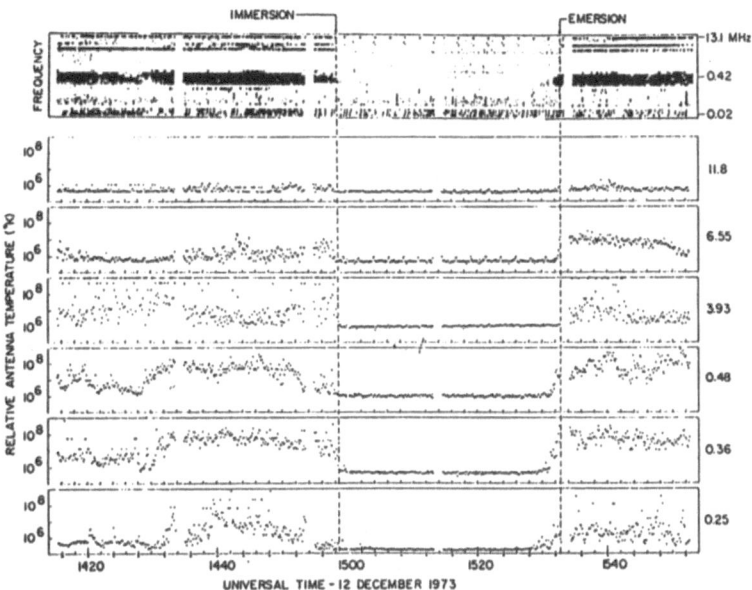

Figure 1. Dynamic spectrum from RAE-2. Note the drop in background levels between 15:00 and 15.30 hrs when the spacecraft was on the far-side of the Moon.

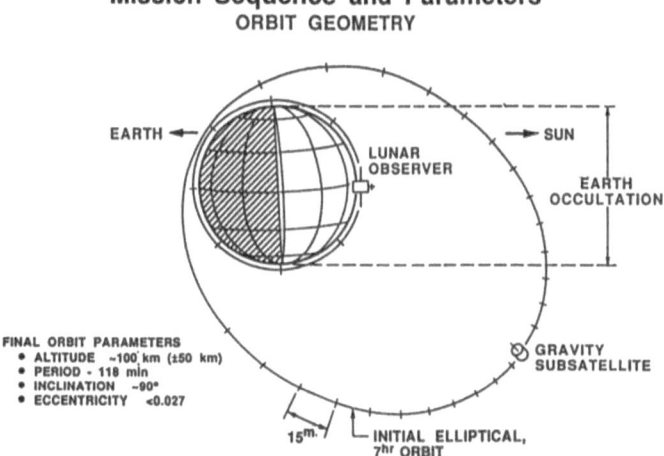

Figure 2. Orbits of the Lunar Observer component satellites.

Observer satellite will present us with a target of opportunity and a "free" ride, thereby placing a low frequency observatory into a particularly attractive space-based environment.

THE LUNAR OBSERVER (LO) MISSION

Preliminary plans call for one, and possibly two (redundant pair), satellites to be placed into a precessing lunar polar orbit. The first observer spacecraft will enter into an initially elliptical orbit with a period of 7 hrs as shown in Fig. 2. After a preliminary low resolution gravity map is completed, a subsatellite, consisting of a simple S-band beacon, will be deployed and the main spacecraft will undergo a circularization burn. The final orbit will place the LO about 100 km above the surface on average. The gravity subsatellite beacon will be used to determine accurate positions for the LO via Doppler shifts so that an accurate gravity map of the far-side can be obtained while the spacecraft is out of direct communication with the Earth.

In the redundant satellite pair concept, a second spacecraft will be placed in a similar high elliptical orbit if the first functions properly. The second LO will likely remain in this orbit.

Thus, there may be up to four separate satellites orbiting the Moon. These satellites offer potential platforms upon which astronomers can place dipole antennas.

LORAE CONCEPTS

We propose placing a crossed-dipole antenna on board the main LO spacecraft. We plan to switch between the antennas to provide a beam with higher directivity than that of a simple dipole.

We propose to operate this experiment between 30 kHz and 30 MHz. Below 1 MHz, where the Galaxy is opaque, observations will focus on studies of the AKR, the Sun, and emissions from Jovian planets. Above 1 MHz, we plan to perform primarily galactic and extragalactic observations. We are suggesting 10 frequency bands with 5 channels each. Breaking the bands into separate channels will allow us to do frequency switching and thus potentially control near-side, Earth-generated interference which is highly variable in time and frequency.

We are proposing three options for LORAE depending upon how the mission evolves -- from a simple dipole experiment on one of the LOs to a multibaseline array.

Option 1 - Single Dipole

This concept is technically simple and is low cost (several x 10^6 dollars). It would be a significant improvement over the RAEs with increased dynamic range and much better calibration.

One of the principal goals of this experiment is to study low frequency interference from the Earth. It is clear that we currently do not understand the nature and magnitude of this interference (e.g., Erickson, 1990). We are in urgent need of more and better data so that we can learn how to control this interference for future arrays in Earth-orbit or on the lunar near-side.

Even with this simple concept, LORAE will be capable of significant science. For example, as discussed by Webber (1990) and Longair (1990), a modest (few tens of degrees) resolution all-sky map at 10 MHz will provide a powerful probe of Galactic cosmic ray electrons when combined with observations from the gamma-ray observatory. At low frequencies, we sample nonthermal emission from a population of electrons at about 200 MeV -- a population of low energy cosmic rays that is not observable from Earth due to solar wind modulation. These electrons are also responsible for producing gamma-rays via relativistic electron bremsstrahlung. Thus, this combination is capable of directly determining the strength of the interstellar magnetic field and electron density as a function of position in the galaxy. We are currently studying dipole configurations that will allow us to achieve the required resolution.

An important additional science goal includes monitoring bursts of low frequency radiation from the Sun and Jovian planets. For example, Saturn's kilometric radiation is strongly controlled by the solar wind. Therefore, by monitoring the flux of Saturn's emissions, we have a highly reliable record of the solar wind conditions at a distance of 10 AU (Desch, 1989). Also, observations of Saturn's bursts or electrostatic discharges near 5 MHz provide extremely accurate information on variations in Saturn's ionosphere -- data that cannot be obtained in any other way.

RAE-2 demonstrated that it is possible to use the limb of the Moon for low frequency lunar occultations. This was done successfully for both Jupiter and Fornax A (M. Kaiser, private communication). The resulting Fresnel diffraction pattern allows one to achieve arcminute one dimensional resolution for strong radio sources (resolution is proportional to $(\lambda/2D)^{1/2}$ where D is the distance to the Moon). As shown in Table 1, we may have the capability of determining flux densities and sizes for hundreds of sources. As a result, accurate low frequency spectra can be determined for strong galactic and extragalactic sources. The shape of the low frequency end of the synchrotron spectrum for AGNs, for example, will provide a useful constraint on theories of particle acceleration and the origin of cosmic rays. Most theories predict approximate low energy cutoffs for the energized particles which could be directly tested with new data from LORAE.

TABLE 1. SINGLE DIPOLE, OCCULTATION EXPERIMENT

	1 MHz	30 MHz
T_{sky} (K)	3×10^7	3×10^4
Effective Aperture (m^2)	10^4	365
Integration Time (sec)	75	14
S_{rms} (Jy)		
1 occultation	3900	50
10 occultations	1230	16
# of detectable sources	>100	300
Width of first null in Fresnel Pattern (arcmin)	17	3

Total Mass = 2.4 kg
Total Power Required = 1.9 W
Data Rates < 1 kbit/sec

Another important goal for LORAE is to measure the lunar ionospheric density. The current best estimate of the plasma frequency near the lunar surface is about 90 kHz (Douglas and Smith, 1985). However, this is not known to better than an order of magnitude. Some models (e.g., Vondrak, 1990) predict that interaction of the solar wind with the lunar surface may result in a much higher ionospheric density (plasma frequency >1 MHz) on the day-side. By observing strong sources as they pass across the lunar limb and scanning through frequency space, LORAE can potentially measure the lunar plasma frequency. This has important consequences for planned low frequency arrays on the surface and for understanding the dynamic behavior of the lunar atmosphere.

Option 2 - A Two-Element Interferometer

If two Lunar Observers are launched, there is a possibility of placing dipole antennas on each satellite. With one spacecraft in low lunar orbit and the other in a higher elliptical orbit, long baseline two-element interferometry could be possible. The maximum baseline might be as long as 8000 km and the shortest could be several hundred meters. Thus, a wide range of one dimensional scale sizes of structure could potentially be studied with this interferometer.

In addition to the goals described above for a single dipole, a two element interferometer would be capable of measuring angular sizes down to arcsecond resolution at 30 MHz as noted in Table 2.

TALE 2. TWO-ELEMENT INTERFEROMETER

	1 MHz	30 MHz
Integration Time (sec)	10^6	10^6
S_{rms} (Jy)	100	1
# of detectable sources	1000	10,000
Diffraction Limit (arcsec)	7	0.2
Effective Resolution (FWHM, arcsec)	900	1

Total Mass = 7 kg
Total Power = 8 W
Data Rates < 1 Mbit/sec

This high resolution lunar interferometer would serve to constrain the spectrum of turbulence in both the interplanetary and the interstellar media through measurements of scintillation of strong extragalactic sources. By having baselines several times the phase coherence length, the apparent scattering disks of individual sources can be accurately mapped, thus providing information on the galactic distribution of interstellar scintillation. This has not been possible at higher frequencies where the source size dominates over scattering. Furthermore, my combining scattering measurements at high and low frequencies, it will be possible to accurately determine the power law index, and to search for the effects of an inner scale on wavenumber (Dennison, 1989).

Some preliminary estimates of the sensitivities and resolutions of this interferometer are given in Table 2. One major problem that must be overcome if this experiment is to be successful concerns the data rates. The Mariner-class buses used for LO are limited in their telemetry rates (< 1 Mbit/sec). At first glance, the single baseline interferometer would take up the full bandwidth of the onboard telemetry system. Thus, either this system must be by-passed (direct downlink to Earth) or the interferometer can be used only part of the time. These options and others are currently under study.

Option 3 - Lunar Orbital Array

There are a total of four separate free-flyers currently planned for the LO project. However, the present design for the gravity subsatellites makes them too small and too specialized to permit low frequency dipoles to be added. But, if the mission (and the subsatellites) is upgraded, then adding dipoles may be possible. Furthermore, the Soviets have been invited to participate in this project and could be responsible for the construction of the subsatellites. The subsatellites would likely be more substantial in this case and could accommodate additional dipoles. Although we rate these two possibilities with relatively low weight, a full array in lunar orbit is not out of the realm of possibility at this early stage.

A Lunar orbital array would have the capability of producing maps at high (arcsec resolution at 30 MHz) resolution. With four or more elements in the array, it will be possible to perform amplitude and phase closure to obtain radio maps. Also, such an array will contain some short spacings that will be necessary at the lower frequencies. Multiple simultaneous baselines could be crucial for fast imaging of rapidly varying sources on the Sun and the Jovian planets. In addition, with an array, redundancy of baselines will reduce errors and increase the reliability of the maps.

A SCIENCE THEME FOR LORAE

A low radio frequency space-based array is uniquely capable of studying basic plasma physics, plasma instabilities, and radiation processes from low density plasmas. A quick survey of potential scientific programs at very low frequencies reveals that nearly all are dominated by these plasma processes. These include comparative studies of the magnetospheres of the Jovian planets; studies of the propagation of cosmic ray electrons in our Galaxy, in external galaxies, in galaxy clusters, and in lobes around active galaxies; investigations of coherent emission processes in planetary magnetospheres, pulsars, and extragalactic sources; and more detailed studies of turbulence in the ISM. Thus, a theme for low frequency science begins to emerge. A low frequency space-based array is best suited to studying a variety of astrophysical plasmas - the dominant component of the Universe - that cannot be sampled in any other way.

SUMMARY

LORAE offers us an opportunity to perform low frequency, space-based astronomy from the lunar near and far sides in the mid-1990's at modest cost. The single dipole experiment is particularly light-weight and low power, and is, thus, a reasonable addition to the lunar geoscience experiments onboard the LO. Even in its most modest form, LORAE will be capable of significant investigations of interference from Earth, and important advances in the areas of solar and planetary radio burst phenomena, propagation of galactic cosmic rays, and the low frequency spectra of strong extragalactic sources. In addition, LORAE will perform an important lunar outpost precursor activity by accurately measuring the density of the lunar ionsphere.

ACKNOWLEDGEMENTS

LORAE was formulated with the participation of a wide variety of individuals. In particular, I would like to thank J. Basart from Iowa State; N. Duric and J. Taylor from U. New Mexico; W. Webber from New Mexico State U.; D. Jones, T. Kuiper, M. Mahoney, and K. Nocks from JPL; M. Kaiser and M. Desch from GSFC; and K. Johnston, K. Weiler, and N. Kassim from NRL. I would also like to thank B. McNamara and W. Sanders for their comments on the text. I am grateful for the advice and support from C. Pilcher, L. Caroff, and R. Stachnik at NASA headquarters. This work was supported by a grant from NASA.

REFERENCES

Basart, J. and Burns, J. 1990, these proceedings.

Burns, J. and Mendell, W., editors, 1988, <u>Future Astronomical Observatories on the Moon</u>, NASA Conference Publication, **#2489**.

Burns, J., Duric, N., Johnson, S. and Taylor, J., editors, 1990, <u>A Lunar Far-Side Very Low Frequency Array</u>, NASA Conference Publication, in press.

Dennison, B. 1989 in <u>A Lunar Far-Side Very Low Frequency Array</u>, NASA Conference Publication #3039, edited by J. Burns, N. Duric, S. Johnson, and J. Taylor, p. 43.

Desch, M. 1989 in <u>A Lunar Far-Side Very Low Frequency Array</u>, NASA Conference Publication #3039, edited by J. Burns, N. Duric, S. Johnson, and J. Taylor, p. 35.

Douglas, J. and Smith, H. 1985 in <u>Lunar Bases and Space Activities of 21st Century</u>, edited by W. Mendell, (LPI: Houston), p. 301.

Erickson, W. 1990, these proceedings.

Kaiser, M. 1990, these proceedings.

Longair, M. 1990, these proceedings.

Vondrak, R. 1990 in <u>Lunar Bases and Space Activities of the 21st Century</u>, Vol. 2, in press.

Webber, W. 1990, these proceedings.

VERY LOW FREQUENCY RADIO ASTRONOMY FROM THE MOON

HARLAN J. SMITH
Astronomy Department, The University of Texas at Austin

I. BACKGROUND

Properly speaking this paper should have been given by Jim Douglas or Jack Burns, who over the last several years have thought by far the most deeply on this question. My involvement, as a non-radio astronomer, came about five years ago when it occurred to me that — in principle at least — one of the simplest possible astronomical instruments for early deployment at a lunar base would be an aperture synthesis interferometer whose small number of elements consisted merely of short dipoles laid down more or less at random from a rover vehicle on the dry dielectric lunar surface. Given the quite low lunar ionospheric density, such a simple low-cost instrument could open up the very poorly known portion of the radio spectrum below 30 MHz, especially the terra virtually incognita below 10 MHz. Douglas did most of the serious work in our joint paper on the subject (1986). Jack Burns, then at the University of New Mexico at Albuquerque, had also been thinking of good uses for lunar radio astronomy. In February of 1988, working especially with Stewart Johnson, he convened a small workshop at Albuquerque with only 15 participants, who however did an extraordinarily thorough job of analyzing the lunar VLFA problem. This workshop has now been published as NASA Conference Publication 3039, 1989. All of the articles in it are thought-provoking, with perhaps special attention to the reviews by several authors (Erickson, Bosart, Kaiser) of what little we now know of VLFA, some of the science remaining to be done in this region (Duric), some of the engineering challenges to be faced in building a lunar far-side array (Johnson), lunar environment and site criteria (Taylor), and the outstanding summary of the problem (Douglas) from which some of the points below are drawn.

After two years it is not clear that much has been added, technically speaking. What has changed is the world of space, reviving hopes and opportunities which several years ago seemed dim and remote. Nevertheless there remain important background questions.

1. Just how important to astronomy is the VLF region? Although radio astronomy began in this spectral region, the steady progression has been one of continuous escalation to higher frequencies, leading over the past decade to near-total abandonment of the VLF region. There are two main reasons for this shift. First, the observing conditions from the ground are truly terrible — almost comparable to trying to do optical astronomy in the daytime — because of problems with manmade RF interference and ionospheric scattering, refraction and absorption all of which make it extraordinarily difficult to do serious work, in particular any high spatial resolution studies. Going to higher frequencies solved all these problems. Secondly, higher frequency generally means younger electrons, closer to the sources of emission, and thus of highest immediate scientific interest.

In spite of the flight to higher frequencies a rich scientific content remains in the VLF region. This meeting brings out in many of its papers some of the work which can be uniquely well done ranging from magnetospheric and plasma physics in the solar system through detailed mapping of the plasma structure of our galaxy out to the trace fossil remnants of earlier activities of other galaxies and quasars, with a harvest of data on pulsars, H II regions supernova remnants, and who knows what else along the way.

A dozen years ago, while writing up the decade requests for the Committee on Space Astronomy and Astrophysics of the Space Science Board, several of us were troubled by the very different orders of magnitude of cost of the many things needing to be done. Concerned that the smaller albeit important ones not be lost in the noise, we suggested a convenient terminology: Major, Moderate, and Modest Missions — those costing more nearly 10^9, 10^8, and 10^7, respectively. With some modifications of terminology and factors of two or three for cost and complexity inflation, this kind of categorization is still being used today.

Where are we by now? Has a compelling case been made for VLFA, to compare with that for HST, AXAF, or SIRTF? The answer is presumably no — at present a Major Mission is out of the question, and I fear that even a Moderate Mission would not have a very strong chance of acceptance. But Modest VLFA Missions and perhaps better yet even Minute (~ 10^6) ones — are probably ready for an early renewed lease on life. Indeed, with new forms of relatively low cost access to space beginning to appear there may be an ecological niche which — should NASA fail to respond — could be filled by Japan, Europe or even countries such as India and China which are now just breaking into the field of space science.

2. *Where is the best place in the long run to do VLFA?* There are only three serious alternatives: relatively near-Earth orbit, lunar orbit, and lunar surface. The overwhelming problem with near-earth orbit lies in the near total wipe-out of frequencies below about 0.7 MHz from interferences by terrestrial auroral emissions and the enormous manmade RF leaking through the ionosphere, especially above 3 MHz. However, as reported at this meeting Weiler and his group are continuing to study a Moderate Mission consisting of perhaps half a dozen satellites in gradually loosening formation, eventually synthesizing an all-sky map at several VLF frequencies. Terrestrial interference is still substantial at the moon, but Burns and colleagues note that various levels of instrumentation of the Lunar Observers (from very Minute on up to Moderate) could give increasingly detailed VLF radio data especially when on the far side of the moon, and by taking advantage of lunar occultations could even give excellent resolution on some sources.

As has been noted for many decades, the moon has one truly unique advantage — its back side is essentially perfectly shielded from the earth, and in principle offers the ideal place for nearly *all* radio astronomy of the future. If this state of radio-quiet can be preserved, radio telescopes of virtually any desired bandwidth, resolution, and frequency (within limits set by solar system and interstellar plasma) can be achieved. With the ever-growing pressure on terrestrial radio frequency usages, it is possible that by a few decades from now most limiting work in radio astronomy will have to plan to migrate to the lunar backside.

However desirable, is this likely to happen? Hence the third question:

3. *Will the moon become accessible on a reasonable timescale?* The answer goes far beyond astronomy. I believe that many factors make lunar bases an effective certainty. These include the human imperative to explore, the rapid progress of science and technology making this a reasonable undertaking, the experience base provided by the Apollo program, the strong US and Soviet manned space programs with Europe and Japan coming on fast, the fact that after Space Stations the moon is so

obviously the next target (this time for a permanent base), the need for a great deal of manned experience at this nearby space outpost before taking the next step to Mars, the prospect for economic return from lunar development, and the competition and/or cooperation with European, Soviet, and especially Japanese space programs. President Bush's speech of July 1989 and subsequent NASA studies specify this as a national goal and give a reasonable timetable of 2002 to 2004 for the first manned return to the moon. When we recall that major space astronomy developments have typically required 15 – 25 years to go from initial planning to fulfillment, now is clearly the time to begin active development of lunar-based astronomy.

II. ADVANTAGES OF THE MOON FOR ASTRONOMY

The Moon offers a unique combination of the most important factors facilitating astronomical observations:

- Ultra-high vacuum (essentially free space, giving full resolving power to optical and radio telescopes)
- Stable solid surface (permits simple, cost-effective telescope mounts and pointing)
- Dark sky (free of airglow and aurorae which afflict terrestrial and near-earth orbits)
- Cold sky (vital for observations in thermal infrared)
- Low gravity (permits lighter, less expensive telescope structures)
- Absence of wind (allows telescopes to consider only static and thermal loads, also ultra-light and simple "domes")
- Rotation (guarantees access to entire sky visible from a given latitude yet allows very long and deep exposures)
- Proximity to Earth (3 second round-trip light-travel time allows real-time control of instruments from, and data transfer back to, Earth)
- Distance from Earth (at 400,000 km, far enough from Earth to be free of most of the many kinds of manmade light, noise, and other pollutions afflicting astronomy)
- Lunar farside (the *only* place in the universe which can never see the Earth, hence ideal for limiting-faintness radio astronomy and SETI work)
- Raw material (inexhaustible supply of shielding and raw material for ceramics, glasses, fibers and metals)
- Landforms (a rich array of landforms able to support gigantic far-future instruments, such as Arecidbo-type radio telescope in lunar craters)
- Room (effectively unlimited area for laying out systems of instruments)
- Support base (common supply available for power, consumables, computer facilities, etc.)
- Access (perhaps the most important of all — the immediate proximity of people to erect, adjust, and maintain the systems).

Free fliers offer some of these advantages as well, but for the foreseeable future they are likely to suffer from their inaccessibility for maintenance and alteration, and from the very high cost of their construction requirements for fully reliable remote operations as well as for pointing accuracy and stability. In contrast — given a lunar base established for whatever reasons — lunar telescopes can be even simpler and lower-cost versions of terrestrial telescopes. Although the initial telescopes will almost certainly work in the UV/optical region, the moon has benefits for VLF systems as well.

III. POSSIBLE STAGES OF DEVELOPMENT OF LUNAR VLF

1. Orbiter. In addition to making some astronomical observations of interest, a lunar orbiter with receivers working at several frequencies in the 0.1 to 10 MHz (and perhaps higher-frequency) region could establish the properties of the poorly known lunar ionosphere, currently believed to have maximum electron densities of about 100 cm^{-3}, giving a plasma frequency of 0.09 MHz. This is an important condition to check, since a very much higher value of plasma frequency would somewhat decrease the utility and importance of lunar-based VLF studies.

2. Test Rig. The initial lunar base will almost certainly be on the near (Earth-facing) side. Terrestrial VLF interference will still be very serious, even though the distance of 400,000 km will reduce the effects by 10^4 over those experienced by Earth-orbiters at 4000 km and by 10^2 over those in geosynchronous orbits.

In the face of this interference what if any VLF activity might be considered for a near-side lunar base? The initial groups of selenauts will be rather busy setting up housekeeping and life support, and may not have much time for science. However, because of its extreme simplicity and negligible cost, a single antenna feeding a frequency-swept receiver at high time resolution should be installed at the very beginning, as a site-test instrument to check for any electromagnetic or plasma anomalies connected with lunar-surface observing, to monitor the levels of earth interference and lunar ionospheric variations, and to record some very low-frequency solar and Jupiter storm activity. The first rover missions going out to significant distances of up to tens of kilometers should install one or more remote units thus creating the first VLF interferometer, with significant scientific value on bright sources.

3. Initial Near-side array. If, as is all too likely, activation of a back-side base appears to be still a decade or longer farther away, serious consideration should be given to constructing a prototype 100-element array over a region of about 20 km. This would give valuable and still low-cost experience in developing the most efficient types of antennas and how to emplace them, in how to power them during lunar night as well as day, in how to link the signals from the individual antennas to the central site, and in how best to process the data at the moon. The array, while often limited especially at its higher frequencies by leakage of terrestrial signals through the earth's ionosphere and below about 0.7 MHz by terrestrial auroral noise, would nevertheless be a powerful instrument for mapping the VLF sky and studying individual sources, and some relatively narrow frequency bands might be almost free of interference. If it became possible by very long rover traverses or remote implantation to install one or more elements at much greater distances in the range of hundreds or thousands of kilometers, and to use a data relay satellite, a very powerful synthesized high-resolution interferometer would then exist. When the Sun and/or Jupiter are above the horizon work on faint sources may often be difficult, but VLF studies especially of solar radio events will track them far out into the corona and may be useful in studying the behavior of plasmas in that region.

4. Back-side Development. If the scientific case for VLFA could be shown to be sufficiently compelling, this alone could drive the initial utilization of the lunar backside within the first few years of the return to the moon. It is more likely that the associated difficulties and costs will delay any major backside activities for many years, until lunar bases, life support, transportation, robotics, and other facilities are highly developed.

In any event, at least three possible modes for setting up a backside VLF system can be imagined. The earliest and perhaps simplest might be feasible if the initial large-scale lunar base is close to the limb, and if efficient long-range and high-capacity rovers are available. In this case a 100 (or more) element array could be laid out on the back side within a one-day travel distance of about 400 km, but still far enough around to keep the Earth far enough below the horizon to avoid significant diffraction of terrestrial radio interference. Depending on the technology of the day, comparison should be made of the economics, reliability, bandwidth, and continuity of operations of a fiberoptics cable from the base to the radio site as opposed to use of one or more lunar orbiters or even relay satellites at Lagrangean points.

However, because of its large smooth floor and convenient overlooking central peak visible from many tens of kilometers around, the optimum location for backside VLF radio astronomy is probably near the center of the backside in the small but unique (on the backside) Mare Tsiolkovsky. Since this is likely to be some 4000 – 6000 "road" kilometers from a near-side base, it would require a major expedition and/or robotic operation to establish the large array using ground transportation. Tradeoffs will need to be studied as to whether a temporarily occupied construction base, served by lunar landers, would be a better way to go. Burns *et al* have outlined such an array.

Finally, since the antennas are so simple and require relatively little in the way of electronics, batteries, and solar power, and since their specific locations are not very important, it might even be cost-effective to eject them one at a time from an orbiter, and to soft-land each with its own small retrorocket. It might even be possible to hard-land them and use the survivors. Metrology of the system would be easy to do using self-calibration on bright sources or an orbiting transmitter. This system would also be relatively easy to expand, although once the antennas no longer have line-of-sight microwave or even laser signal-transmission to a common point, use of a more complicated and expensive satellite system would be needed.

By four or five decades from now, when the moon has been extensively developed with a variety of bases, we may expect a VLF system to have been built which effectively makes use of the entire lunar backside giving arcsecond imagery at 20 MHz and one-third arcminute imagery at 1 MHz. These values would push on the threshold of scattering ("seeing") by electron irregularities in the interplanetary and interstellar medium

IV. SUMMARY

Lunar development is virtually certain to occur in the next 10 – 20 years. Very low frequency radio astronomy observations can and should begin with the first lander. A modest yet powerful near-side array should be built if backside development will be substantially delayed. A major backside array should be constructed when technology and funding permit. Ultimately the full dimensions of the radio-quiet backside of this splendid natural spacecraft will be exploited to give several arcsecond resolution at 10 – 20 MHz and fractional arcminute resolution below 1 MHz. It is already time to begin long-range planning for these developments, and in particular to preserve through internationally recognized bodies as much as possible of the radio-quiet status of the lunar backside.

AN EVOLUTIONARY SEQUENCE OF LOW FREQUENCY

RADIO ASTRONOMY MISSIONS

Dayton L. Jones
Jet Propulsion Laboratory
California Institute of Technology
Pasadena, CA. 91109

Many concepts for space-based low frequency radio astronomy missions are being developed, ranging from simple single-satellite experiments to large arrays on the far side of the Moon. Each concept involves a different tradeoff between the range of scientific questions it can answer and the technical complexity of the experiment. Since complexity largely determines the development time, risk, launch vehicle requirements, cost, and probability of approval, it is important to see where the ability to expand the scientific return justifies a major increase in complexity. An evolutionary series of increasingly capable missions, similar to the series of missions for infrared or X-ray astronomy, is advocated. These would range from inexpensive "piggy-back" experiments on near-future missions to a dedicated low frequency array in Earth orbit (or possibly on the lunar nearside) and eventually to an array on the lunar farside.

I. Ground vs. Space

The major reason for planning space-based low frequency (LF) astronomy missions is to get above the Earth's ionosphere. The frequency below which useful observations are not possible from the ground varies greatly with time and location, but in general it is extremely difficult to observe below ~ 10 MHz. Very useful observations could be made above 10 MHz by an orbiting or lunar-based low frequency interferometer array, but the strongest justification for such an array would be observations at lower frequencies.

The scientific questions which can be answered only with high angular resolution observations below 10 MHz include:

- the distribution of diffuse ionized gas and low energy cosmic rays in our galaxy.
- the physical conditions (magnetic field strengths, electron densities, radiative lifetimes) in very extended radio sources.
- the importance of coherent radio emission processes.
- the evolution of solar radio bursts through the unobserved frequency range between ground-based and current space-based instruments ($\sim 2 - 20$ MHz).

Figure 1 shows the frequency range of interest, along with the observational capabilities of some existing and proposed instruments.

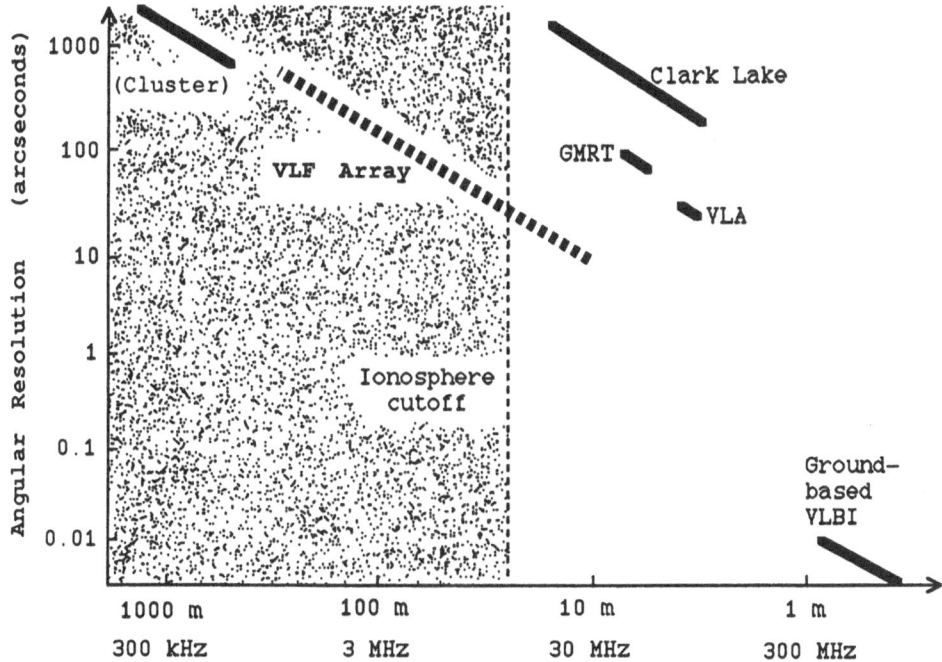

Figure 1. The angular resolution of several existing and proposed radio interferometer arrays as a function of frequency. The ionosphere cutoff, below which high resolution observations are impossible from the ground, varies with time. The GMRT is a large array under construction in India, and the VLA curve is for 75 MHz operation of the Very Large Array (currently installed on only a few antennas, but proposed for the entire array).

The Cluster array shown in Figure 1 is intended to study plasma processes in the Earth's magnetosphere, and will observe at lower frequencies than the other space arrays considered here.

II. Observational Capabilities

The capabilities required for various types of observing are listed below. They range from simple monitoring of natural and man-made radio interference levels (important for the design of dedicated LF arrays) to accurate high dynamic range imaging of weak or rapidly variable sources.

Observation	Instrument Requirement
RFI data	Spectrum analyzer on single spacecraft
Lunar occultations	Receiver/antenna on single (high) lunar orbiter
Angular size of strong sources	Two spacecraft (variable spacing)
Imaging of strong sources	Three spacecraft (closure phases but not amps)
High dynamic range images	Four or more spacecraft, wide range of spacings
Weak source imaging	Long integration times, good u, v coverage
Map entire sky	Long integreation time, very good u, v coverage
Image rapidly varying sources	Many simultaneous baselines (snapshots)

The following sections consider several LF missions in order of increasing complexity.

III.   Single-Spacecraft Missions

Any high Earth orbiter or a lunar orbiter could provide valuable information on the strength and time variability of RFI throughout the frequency range of interest for future low frequency arrays. A spacecraft in high lunar orbit could also be used for radio occultation measurements, providing data on the sizes and positions of strong sources. However, a large number of occultations of the same source will be needed to provide useful sensitivity. The second Radio Astromomy Explorer mission used this technique nearly two decades ago. One or more of the Lunar Observer spacecraft may be placed in a suitable lunar orbit. The angular resolution of lunar occultation observations is $\sim (\lambda/2D)^{1/2}$, where D is the distance from the Moon.

IV.   Single-Baseline Interferometers

Single-baseline VLF interferometry has already been successfully demonstrated with the ISEE-1 and ISEE-2 spacecraft at ≈ 250 kHz. Fringes were detected from the AKR emission region, allowing an upper limit for the size of the emitting region to be determined.

The Lunar Observer (LO) mission is currently planned to have two main spacecraft in lunar orbit (each of which will deploy a small sub-satellite). It may be possible to place low frequency receivers and antennae on each main spacecraft. This would provide an opportunity for a near-term, inexpensive radio interferometry experiment, including some obeservations from the radio quiet farside of the Moon. The technical issues involved in this experiment, such as the low bandwidth of the data system on the LO spacecraft and the non-optimal orbits they will be in, are being studied. Calibration of data from a single-baseline VLBI experiment is likely to be very difficult.

V.   Four-Telescope Arrays

Three-telescope arrays have been skipped since they are not being actively proposed or studied. The advantages of having four spacecraft instead of three (mainly the ability measure closure amplitudes) are worth the fractional increase in complexity.

There are at least three separate plans for 4-telescope LF arrays in space:

1. The ESA Cluster mission will place 4 spacecraft in elliptical polar Earth orbits. VLBI observations at frequencies up to $\approx$ 2 MHz will be used to study terrestrial plasma emission, and attempts will be made to detect more distant sources.

2. As mentioned in the previous section, the Lunar Observer mission is expected to include two small sub-satellites in addition to the two main spacecraft. Low frequency receivers could be added to all four spacecraft, although there are many technical problems associated with this concept. It remains to be seen if they can be overcome.

3. The Low Frequency Space Array (LFSA) is the only dedicated mission for low frequency VLBI from space. It would be an extremely valuable mission, having much greater sensitivity, (u, v) coverage, and flexibility than any "piggy-back" experiment. As noted below, this mission may well include more than four spacecraft, but is listed here because four is the minimum number of spacecraft which have been considered.

VI.   Large ($>$ 4 Telescope) Arrays

There are at least 5 concepts being studied which would provide more than 6 simultaneous baselines. Three of them are expanded versions of the missions listed in section V.

1. The ESA Polar spacecraft is expected to fly during the same time period as Cluster, opening the possibility of using all five spacecraft to provide a 10-baseline array (at frequencies below about 2 MHz) when over the polar region.

2. There has been some discussion of deploying additional sub-satellites into lunar orbit from the Lunar Observer spacecraft. Detailed study of this option has not yet begun.

3. The LFSA mission is currently planned to have more than four spacecraft in Earth orbit, providing improved imaging (especially of rapidly varying sources).

4. Lunar-based arrays are being studied more actively as a result of the recently announced "Lunar Initiative". One concept for a nearside array which could be deployed early in the development of a lunar base is described in section VII.

5. A large VLF array on the radio-quiet lunar farside is the ultimate instrument for low frequency radio astronomy. However, such an array is decades away. Nevertheless, farside array concepts are being discussed and will be covered in section VIII.

VII. Lunar Nearside Arrays

The Lunar Initiative has increased the relevance of lunar-based astronomical instruments. There is, of course, no guarantee that this initiative will receive the funding needed for rapid progress, but if it does then the development of a more efficient system for transporting mass from Earth to the lunar surface will likely be a high priority. This could result in it being less expensive to deploy a low frequency array on the Moon (see Figure 2) than in high Earth orbit. The technical simplicity of a low frequency array (compared to proposed lunar-based interferometers for higher frequencies) would make deployment possible very early in the establishment of a lunar base.

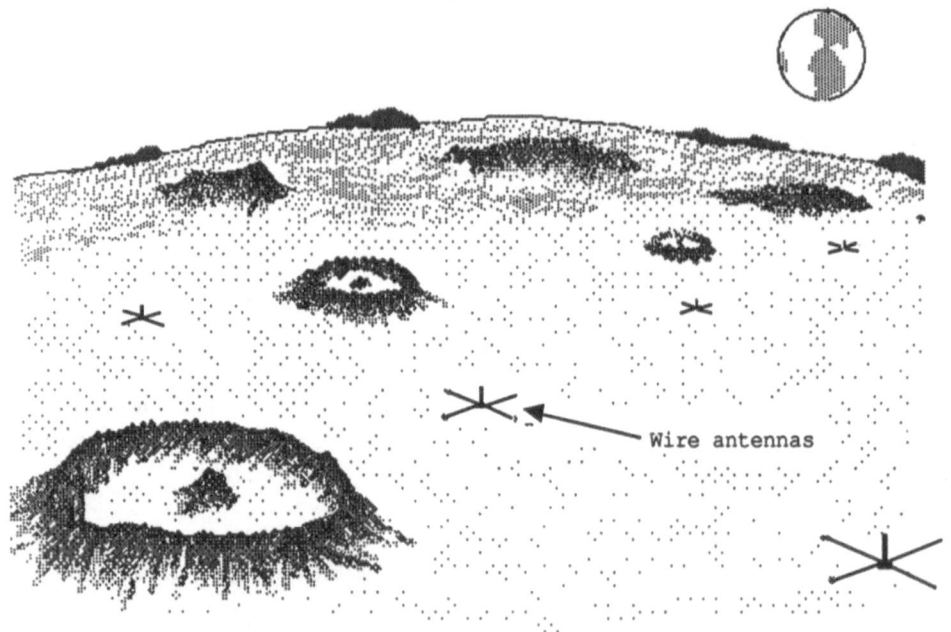

Figure 2. One concept for a nearside array. Each element is composed of crossed wire dipoles on the lunar surface and a central mast to support a high frequency data relay antenna (see figure 3).

In this version, data from each antenna element is relayed (at some high radio frequency) by neighboring elements to a central location. Other options include the use of optical fibers for data transmission, direct transmission from each antenna element to a mountain-top site visible from the whole array, and direct data transmission to Earth.

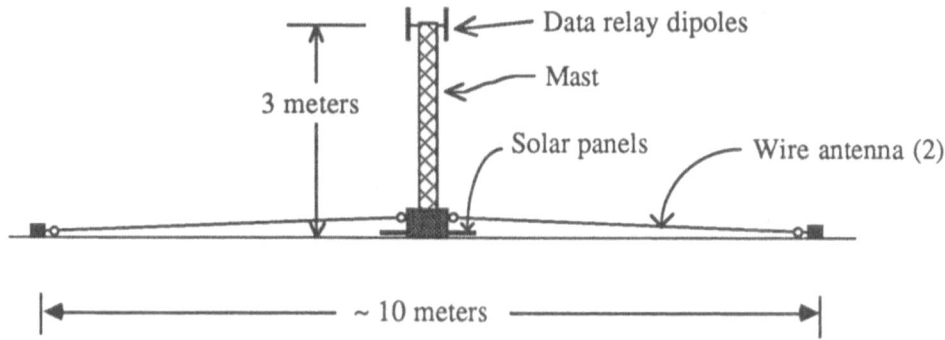

Figure 3. Possible design of individual antenna elements. The wire dipole antennas are shown above the lunar surface for clarity, but they would actually be placed directly on the surface.

A lunar-based array has some obvious advantages and disadvantages compared to an Earth-orbiting array. The main advantages are the greater distance from Earth (lower RFI levels) and the stable, known array geometry. The main disadvantage is that without a yet-to-be-developed transportation infrastructure it will be much more difficult and expensive to place an array on the lunar surface than to deploy one in Earth orbit.

VIII.   Lunar Farside Arrays

The far side of the Moon is the only location anywhere near the Earth which is shielded from strong natural and man-made terrestrial radio interference. This is the one big advantage of a farside array. The most sensitive radio observations will require a site on the lunar farside, but even with the Lunar Initiative it will be decades before any major instruments are deployed there. The promise of a farside array, although scientifically exciting, should not deter the development of Earth orbiting and/or lunar nearside arrays which could answer many important scientific questions during the next 5-15 years.

IX.   An Evolutionary Approach

The various experiments and missions described above can be grouped into three general catagories, as shown in Figure 4.

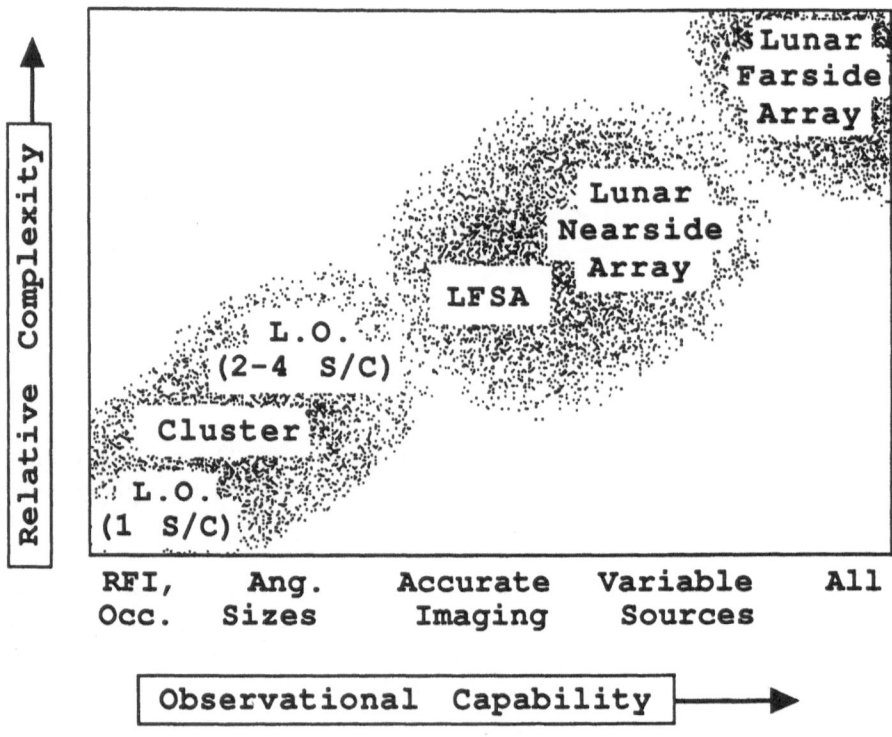

Figure 4. Continuum of VLF array complexity and capabilities. The three shaded areas enclose "piggy-back" experiments on relatively near-term missions, dedicated VLF imaging missions which could fly in the late 1990s or early 2000s, and far-future missions (later than 2010).

Experiments or missions in each of these catagories will provide qualitatively different scientific returns. Each is a reasonable tradeoff between scientific potential and complexity (cost), and thus missions in each catagory should be pursued.

X. Summary

A progression from simple, near-term experiments to Explorer-class dedicated missions to a large lunar farside array is a natural way to develop the observational capabilities needed for radio astronomy at low frequencies. This continuous increase in mission capabilities and scientific return is similar to the progression of infrared and X-ray missions.

In the infrared, we have the following sequence:

1. balloons, sounding rockets, and the Kuiper Airborne Observatory.
2. an Explorer class mission (IRAS).
3. a Great Observatory class mission (SIRTF).
   and similarly for X-ray missions:
1. balloons and sounding rockets.
2. Uhuru.
3. HEAO.
4. AXAF.

It can be hoped that the initial "biggy-back" VLBI experiments on the Cluster and (possibly) Lunar Observer missions will be followed by a dedicated LF interferometry mission. The greatly increased observational capabilities of a mission like the LFSA (or a lunar array if that becomes practical) would open up a whole new range of scientific opportunities. It is an exciting prospect.

This work was carried out at the Jet Propulsion Laboratory, California Institute of Technology, under contract with the National Aeronautics and Space Administration.

HECTOMETER AND KILOMETER WAVELENGTH RADIO ASTRONOMY

Grote Reber
Michael Street
Bothwell, Tasmania
Australia 7030

INTRODUCTION

The two main reasons for doing low frequency radio astronomy observations are:

A. It provides a technical challenge both as to equipment and to the understanding and use of the earth's atmosphere.

B. The radio sky at hectometer and kilometer wavelengths is very different from the radio sky at meter wavelengths and shorter. I consider this one first.

SCIENTIFIC MOTIVATIONS

The radio sky is similar to the optical day sky; bright all over with brightness maxima near the galactic poles. The <u>apparent</u> temperature rises from 6×10^4 degrees at 19.7 MHz (Reber 1964, see Fig. 1 p. 258) to 10^7 degrees at 1.6 MHz (Ellis and Mendillo 1987) at the south galactic pole. The nomenclature is bad. There is nothing out there with any such temperature. The intensity merely appears that way if there were. The source of this radiation is probably encounters between free electrons and protons in intergalactic space (Reber 1968). I call them free-free transitions. Others refer to it as bremsstrahlung or braking radiation. However the matter, it is clear the phenomena are not thermal.

According to thermodynamics people, the universe is running down. However the universe has been around a long time. It seems probable some phenomenon, not thermodynamic, is winding it up. This phenomenon is probably associated with the high intensity background mentioned above and is probably electro-dynamic. I have discussed this in some detail (Reber 1964, see p. 262). During 1952 I built a small electrostatic motor. It ran nicely when supplied with energy having a continuum of periods. No non-linear devices such as rectifiers were used (Reber 1964, see p. 262). This is one aspect of cosmology which deserves more investigation. In fact, the whole expanding universe theme is peppered with difficulties (Reeves 1989). These difficulties seem to say there is a fundamental error of interpretation of the evidence. For example, that the red-shifts have nothing to do with relative motion (Reber 1986). In my opinion, the exciting frontier of future radio astronomy is in the nearly unexplored region of waves longer than 100 meters.

TECHNICAL CONSIDERATIONS

Being an observer and technician, I am most interested in the equipment. I will concede that observations above the ionosphere are desirable. However the RAE-1 satellite (Weber, Alexander, and Stone 1971) demonstrated this is no panacea. The ionosphere is leaky and did not shield the satellite from man-made emissions. These are of great intensity and completely swamped the radio astronomy receivers. Accordingly, satellites circling the earth are out. The difficulty will probably extend as far as the moon and beyond. To secure really quiet location to do low frequency radio astronomy, the observer must be on the back of the moon (Alexander et al. 1975). Such may eventuate by the middle of the next century, long after my time.

The possibilities here on the surface of the earth have not been properly exploited, much less exhausted. During northern winter 1986/87 exploratory observations were made at 2.1 MHz using an old antenna array just west of Ottawa. The north galactic pole has the same high intensity as the south galactic pole. There is a 30% decrease of intensity when the antenna beam crosses the plane of the Milky Way north of Cygnus. These results are reassuring. Then an ice storm brought most of the antennas down. Solar activity was rising, time had run out, and nothing more was done.

Operating an ionosonde is a popular scientific sport. It is being indulged in at scores of places around the earth. This data is analyzed for the maximum critical frequency. This is important to predict high frequency radio communications. I've examined some of this data for low critical frequency, particularly from Ottawa, Canada, Boulder, Colorado, and Wallops, Virginia. During the solar activity minimums of the 1970s and the 1980s, the critical frequency drops below 2.0 MHz at Ottawa 5.5 times as often compared to Boulder and 7.1 times compared to Wallops. This confirms the Alouette satellite data. There is a marked valley in the F region electron density near 45^0 latitude in southeastern Canada. The ionosondes at Winnipeg and St.

Johns have been shut down. I am requesting copies of hourly values for $f_0 F_2$ for the appropriate years, but so far nothing has come to hand.

A similar valley crosses over Tasmania. It was used to map the southern sky during solar minimum of the mid 1960s (Reber 1968). These showed that the ionospheric hole at 2.1 MHz opened for at least a few hours every winter night. The hole stayed closed all night for about three weeks near summer solstice. Some winter nights the hole was open 10 and even 12 hours. Beautiful clear charts showing great detail were secured. The absorption in the ionosphere was sensibly zero. On clear nights the traces would reproduce within the accuracy of the measurement, about two percent. The single beam could be set anywhere in North-Zenith-South plane from the celestial equator to south celestial pole. A map of the southern sky was developed (Reber 1968). The antennas were broadbanded from 1850 to 2450 KHz. By modern standards the receiver is archaic, using 1.4 volt tubes. However the battery supply provided excellent gain stability.

The main equipment deficiency was having only one beam. It could be changed in direction by a complex system of taps on feed lines. These took the better part of half a day to change. When a good night happened, only one trace was secured. To take advantage of these occasions, a picket fence of beams along the meridian from equator to pole should be available. Such would greatly speed up a sky survey. Accordingly a new set of receivers and fixed phase shifters have been constructed. These provide 25 beams. No outside or inside adjustments are needed. Everything is fixed. The antenna looks at 25 different directions simultaneously. However this arrangement cannot be used for transmitting, as was possible with the original setup. The main thing needed now is an antenna array of large dimensions, at an optimum place to be prepared for the solar activity minimum of the mid 1990s. I've examined swinging the beam east/west and found that this can be done. In fact a single beam can be made to track an object in the sky at any place within some 50^0 of the Zenith. As might be expected, the complexity goes up by leaps and bounds.

CONCLUSIONS

All in all, I recommend more attention be paid to ground based installations. The satellite and moon stuff is very expensive, long drawn out, and of dubious performance. Just pie in the sky.

REFERENCES

Alexander, J.K., Kaiser, M.L., Novaco, J.C., Grena, F.R., and Weber, R.R. 1975, Astr. Astr., **40**, 365.

Ellis, G.R.A. and Mendillo, M. 1987, Austr. J. Phys., **40**, 707.

Reber, G. 1964, IEEE Transactions on Military Electronics, July-October, p. 258.

Reber, G. 1968, J. Franklin Inst., **285**, 1.

Reber, G. 1986, IEEE Trans. on Plasma Science, December, p. 678.

Reeves, H. 1989, J. Royal Astr. Soc of Can., **83**, 223.

Weber, R.R., Alexander, J.K., and Stone, R.G. 1971, Radio Science, **6**, 1085.

A SIMPLE LOW-COST ARRAY ON THE LUNAR NEAR-SIDE FOR THE EARLY LUNAR EXPEDITIONS

T.B.H. Kuiper, D.L. Jones, M.J. Mahoney, R.L. Preston
Jet Propulsion Laboratory
California Institute of Technology

Last September, JPL was charged with reporting to NASA HQ on science that can be done as a part of a return to the Moon. As is typical, for years and years nobody listens when you're trying to get a study going, and then all of a sudden it has to be done immediately. So, while these VU-graphs look very spiffy and suggest that this project has been thought about a lot, in truth, all of this comes out of some discussions that Dayton, Mike Mahoney, Bob Preston and I had over lunch and similar informal occasions. Hopefully these have smoothed the roughest edges off a really preliminary idea. I then followed up with a few phone calls to some engineers at JPL. It's one of the advantages of being at a place like JPL, that you can pick up the phone, ask technical questions, and get answers quickly. There is not a lot of work in this, and I want you to understand that.

I put this [picture of the Clark Lake TPT] up for a number of reasons. One, is to establish my credentials. In so far as I am known in astronomy I'm known as a cm and mm wave astronomer studying interstellar molecules. But, this is where I did my thesis. This array was going up when I was working on my thesis, and here one can vaguely make out one of the elements of the Log-Periodic Array, which I used to do a thesis on Type III bursts. I share that distinction with Tom Gergely, who is hiding in the back of the room over there.

I also show it for another reason, to tell you that a low-frequency array on the Moon is not something that we invented since the President said we are going back. I presented an idea to an internal JPL conference four years ago, just about exactly today, for a concept based on the TPT. At the same time I presented an idea for an array of simple cheap satellite orbiting the Earth. So these two ideas are contemporaneous and pretty old. Other people have thought about these things on that sort of time scale and longer.

Let me now move immediately to the concept (Table I). I'll come back to this one in a second. I'll just give you a real quick visual idea (Figure 1). A T is our current preference, but it's still being kicked about, and does change from one lunch time to the next. This is a schematic of each single station consisting of some dipoles laid orthogonally over the ground, and a mast for the telemetry.

I'll describe that in more detail with the help of this viewgraph (Table II). To show you how the thinking went, we thought what sort of structure could you have that was very simple and very light, because you don't want to carry a lot of mass up, in order to carry the signal from a particular antenna eventually to the center of the array, A phone call to a structural engineer at JPL suggested that using a "Jack in the box" antenna we might be able to erect something which is three meters tall. Then taking the curvature of the Moon into account and adding the requirement of at least being able to see the second antenna down the line, so that if you loose one antenna you don't loose the telemetry link to the center, that would result in a spacing between antennas of six km. We thought something like twenty antennas would be a reasonable number and that limits the size of the array to being a total of 70 km along the long arm.

Table I - Mission Concept

Nineteen stations are laid in a T, with three arms each 36 km long (stations separated by 6 km).

Each station has two 10-m long orthogonal dipoles, each feeding a receiver like a commercial pocket digital shortwave radio and digitized by circuitry used in digital audio tape recorders.

A 3-m spring-loaded mast with a small UHF array at the top receives timing signals and tuning commands, and sends out the digitized data streams, using commercial transceiver technology. Each station can relay signals from other stations. Each station can "see" two other stations in both the inward and outward directions, providing relay redundancy.

Each station has 0.25 m² of solar arrays and 20 kg of lithium batteries. Power is collected during the lunar day. Observations are made during lunar night, when radio interference from the Earth is low

At the array center, either (1) a digital correlator processes the signals from each antenna pair to compact the data into a 32 kb/s channel, or, (2) a 10 cm telescope at 100 mW laser sends back 400 MB/s of raw data to Earth

Performance: Up to 22 kHz bandwidth tunable from 150 to 30000 kHz

Angular resolution of 18' at 1 MHz, 6' at 3 MHz, 2' at 10 MHz

Sensitivity is limited by radio emission from the Galaxy, not electronics technology

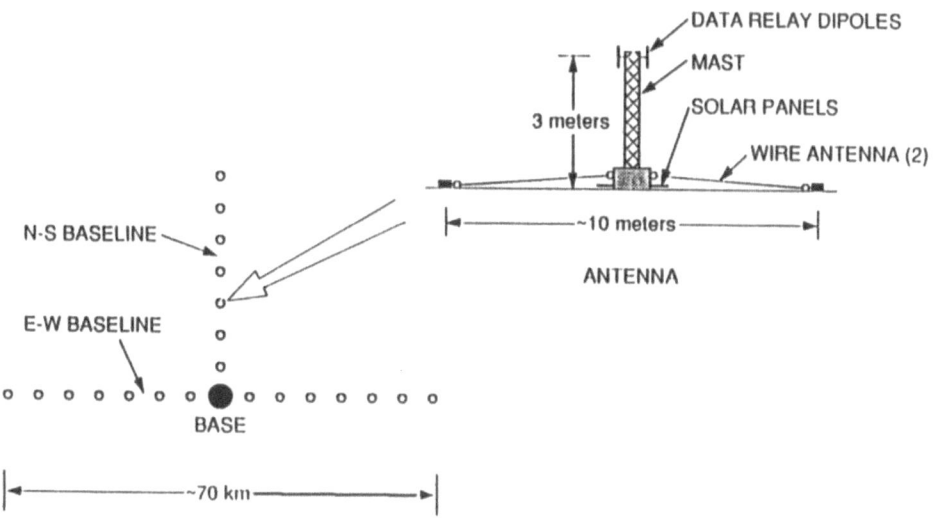

Figure 1 - System Configuration

The receiver is very simple. I have a little Sony digital shortwave radio. It's about the size of a pocket book, costs about $300.00, is digitally tuned, crystal controlled, and goes from 150 kHz to 30 MHz. Most of it consists of the speaker, the battery, and the keyboard and the LED display. The amount of electronics is insignificant. And yet, that sort of electronics has all the sensitivity that you can use because of the intense galactic

Table II - Principal Subsystems and Status

SUBSYSTEM	FUNCTION	TECHNOLOGY STATUS
DEPLOYABLE ANTENNAS	RECEIVE RADIO WAVES	LOW TECHNOLOGY; SPECIFIC MECHANISM TO BE SELECTED; AUTOMATION AND ROBOTIC TECHNOLOGY NEEDED FOR DEPLOYMENT
DEPLOYABLE MAST	TRANSMIT, RECEIVE INTERSTATION TELEMETRY	STANDARD SPACE TELEMETRY HARDWARE ("JACK-IN-THE BOX")
RECEIVERS	TUNE IN AND AMPLIFY SPECIFIC FREQUENCY, AND CONVERT TO AUDIO FREQUENCY	COMMERCIAL
DIGITIZERS	CONVERT AUDIO FREQUENCY TO DIGITAL DATA	COMMERCIAL
INTERSTATION TELEMETRY TRANSCEIVERS	RECEIVE COMMAND AND TIMING SIGNALS, TRANSMIT DATA, RELAY SIGNALS AND DATA FROM OTHER STATIONS	COMMERCIAL
STATION POWER SUBSYSTEM	PROVIDE CONTINUOUS 5W AT 15V DURING LUNAR NIGHT	STANDARD SPACE HARDWARE
CENTRAL STATION CORRELATOR OR	CORRELATE SIGNALS IN REAL TIME FOR ALL 171 BASELINES	CURRENT POWER CONSUMPTION IS HIGH; DRAMATIC REDUCTIONS EXPECTED IN <5 YEARS
LASER TELEMETRY TO EARTH	SEND ALL RAW DATA TO EARTH	EXISTING TECHNOLOGY; SPECIFIC HARDWARE FOR USE ON MOON NEEDS TO BE DEVELOPED

background emission. So that's the electronics that we are talking about. There's no mass in that electronics.

Regarding the telemetry, there are hand-held transceivers and telephones and so forth. They have all the technology needed to bring clock signals from the center out to each element and to take the digitized data back to the center. That's insignificant in electronics. Most of the mass, in fact, is in the batteries because, as you just saw in Erickson's paper, you can only expect observe on the nearside when it is daytime on the Earth, and that's nighttime on the Moon. So, you've got to store power during the daytime and expend it during the nighttime. That tells you how much power you need to store. We are anticipating about a 5 W rate of expenditure. Taking the length of the lunar day and night into account, that tells you how many batteries you need, and the size of the solar arrays. So those are the reasons for the specific numbers.

Originally, we thought about having a correlator at the center but then Dayton came back with the idea of just sending all the raw data back, and that's really quite preferable for a number of reasons. I think it's simpler technically. I think that it involves --at least with the technology we have available today-- less mass to carry to the Moon, and you have the benefit of being able to use the latest technology on the Earth, whereas if you put your correlator on the Moon you have to freeze your technology, and that particular technology is as you well know is changing by the year. So that's the preferred option. But we do need to look more closely at the details of actually having a little telescope like that and what it requires in terms of pointing and so forth. It doesn't seem like a formidable problem, but it is one that we need to look at a little more closely.

So then, in summary (Figure 1), we've got the mast with the little relay dipoles at the top, the batteries in here, little solar panels which flop out and then the wire antennas. We envision that a lunar rover, either manned or automated, drives along the lunar surface, puts a box like this down and drives away. A little time later, little weights shoot out from the side carrying wires dropping them on the surface, the solar panels fall down, the mast comes up, and the station is in place.

Assuming that we've taken a reasonable number of (specifically nineteen) antennas, these are some of the specifics of what you will achieve in terms of resolution (Table II). Incidentally, the 22 kHz bandwidth is also something which comes from commercial technology. That's the bandwidth that's used in digital tape-recorders, and CD's. And so again, the chips are available, including not only digitizing but also encoding for redundancy so that you have some protection in case of data loss. All of that stuff exists now commercially for a very low price.

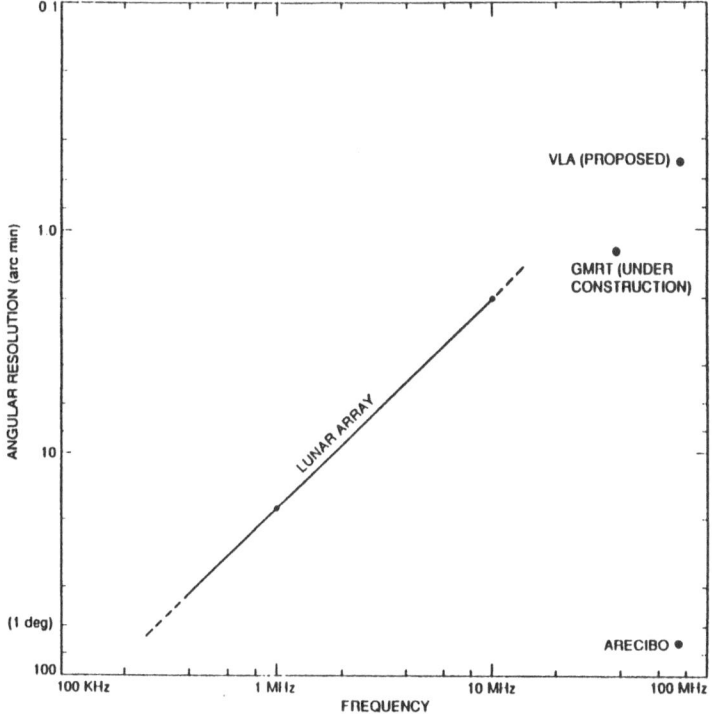

Figure 2 - Angular Resolution *vs* Frequency

The resolution is shown graphically in Figure 2. The 70 km end-to-end array fits in nicely with arrays at higher frequencies.

Of course, we'd like a bigger array, but the idea of this mission is that when we go back to the Moon, we can immediately begin to do useful radio astronomy with something which is really no bigger than a rover loaded down with twenty boxes about this size.

There is really no technology to develop. There are some specific gadgets that we need to design and probably it makes sense to actually build them and test them. But, unlike so much other astronomy, there's really no technology development as such. In fact, some of the things are commercial and I would imagine that a useful way to actually think about designing the electronics would be to get on an airplane and go over to Sony and talk to their engineers because I think they have all the chips sitting on the shelf.

Here is a summary of the costs and weights (Table III), and again, they're very preliminary. In some cases we're dreadfully honest: we say that it's only a guess. In the case where we have existing commercial technology we use a factor of twenty to give us an idea what the real cost might be. In other cases we called up an engineer at JPL and asked him to make his best guess. So, you can put your own factor on that. If you believe these numbers the total cost of the mission is about $20M. If we are wrong, maybe it's $40M. It's still, I think, very cheap compared to what it normally cost to do missions in space.

Table III - Mass and Cost Estimates

SUBSYSTEM	BASIS FOR MASS	BASIS FOR COST
DEPLOYABLE ANTENNAS	0.5 kg: PRELIMINARY DESIGN	$100K: ENGINEERING ESTIMATE
DEPLOYABLE MAST	0.5 kg: ESTIMATE	$10K: "JACK-IN-THE BOX" ANTENNA
RECEIVERS	0.3 kg: SONY ICF-2002	$10K: SONY ICF-2002 COSTS $250
DIGITIZERS	0.2 kg: DIGITAL AUDIO TAPE RECORDER	$10K: 20x COMMERCIAL EQUIVALENT
INTERSTATION TELEMETRY TRANSCEIVERS	0.5 kg: HANDHELD TRANSCEIVERS	$10K: 20x COMMERCIAL EQUIVALENT
STATION POWER SUBSYSTEM	20 kg: EXISTING SPACE HARDWARE	$30K: GUESS
CENTRAL STATION CORRELATOR, OR,		$10M: 10x COST OF VLA CORRELATOR
LASER TELEMETRY TO EARTH	35 kg: GUESS	$1M: GUESS

We see this as a step in an evolving program. One could imagine a program that looks like this (Figure 3). We believe it's extremely valuable to get more information on the RFI environment. How well can you use, for example, frequency agility, the idea that you can hop around and choose your best frequency in almost real time? That could be studied with a receiver on Lunar Observer. At the same time, I don't think that we should deploy something without having tested it

```
    LUNAR OBSERVER                SCALED TEST ARRAY
 low frequency receiver           20-40 Mhz on Earth
    RFI/AKR studies                deployment tests
  large scale mapping              data management
   lunar occultations              RFI cancellation

              LUNAR NEARSIDE COMPACT ARRAY
           verify deployment and data handling
                mapping during lunar night
              AKR, solar bursts during day

                    FARSIDE ARRAY
                   300 km diameter
                improved antenna design
                  no terrestrial noise
```

Figure 3 - A Possible Program

because the cost of testing it is so low. We should build something and take it out to Death Valley, or back to Clark Lake, and set it up to make sure that we've got the deployment technology down, and that we understand the details of the data management and RFI cancellation. Then, when we go to the Moon, we can do the mission that I described as a first step to going to the backside.

Incidentally, I should point out, and you probably infer that from things you've heard earlier today, that the nearside array would not be abandoned. Indeed, it would probably be expanded by people who are interested in AKR and other terrestrial phenomena. So I think it would continue to be a useful instrument.

Lastly I'll iterate what somebody else has said, and Tom Gergely will address this later. We should not expect that the back side of the Moon is going to be radio quiet environment, unless we do things now to ensure that it is.

Burns: Tom, you show the dipoles having popped out of the side being elevated above the surface. Is there a reason why they could not lie right on the surface?

Kuiper: No. The reason it looks like that is because that's what came out of Dayton's Mac. In actual fact, the wires would lie on the surface.

Stone: Can I ask you where you get your cost estimate for [the mission]?

Kuiper: In the case where you saw twenty times the commercial equivalent, that assumes the development needed to make commercial hardware into space hardware. In other cases, I asked an engineer at JPL what would it cost to build something like that for this application. It's very nebulous and very rough.

Voice: Does the cost include the actual operational requirements for placing placing the antennas in the array configuration?

Kuiper: No. It doesn't include, for instance, the rover which you would need to deploy it.

A VERY LOW FREQUENCY ARRAY FOR THE LUNAR FAR-SIDE

John P. Basart
Electrical Engineering Dept., Iowa State University
Ames, IA 50011

Jack O. Burns
Astronomy Dept., New Mexico State University
Las Cruces, NM 88003

Introduction

The backside of the moon uniquely provides shielding against man-made interference and the earth's auroral kilometric radiation, while also providing lots of real estate for a low-frequency radio telescope (Douglas and Smith, 1985). With an increased national interest in returning to the moon, it is appropriate to contemplate various technical aspects of a low-frequency lunar radio telescope.

We discuss an initial design for a low-frequency array for the lunar far-side (see Basart and Burns, 1989 for more details). Proper-ties of the telescope that we assume are 1) a frequency range of 1 to 30 MHz, 2) at least four bands of operation (e.g, 1, 3, 10, and 30 MHz), 3) linear polarization (at least initially), and 4) a maximum bandwidth of 5 MHz.

Critical factors in the design of a lunar telescope which differ from an earth-based array are the mass of material transported to the moon, deployment of the array, and electrical power at each array element. The irregular terrain on the far side of the moon makes deployment difficult, especially if deployment is executed by a remotely controlled vehicle. Thus, array elements must be physically simple to deploy and require minimal maintenance. The long lunar night means that nonsolar power must be used for two weeks, and the low mass requirement means that we must avoid stringing copper wires over many miles of the lunar surface to connect the array elements.

Factors to be considered in the far-side array that are directly related to the type of astronomy that can be done are: 1) the type of array (correlation or phased), 2) the configuration of the array, 3) the array elements, 4) the receivers, and 5) the communication network that connects the array elements to a central processing location.

Type of Array

Two types of arrays considered for a lunar observatory are a
correlation array and a phased array. In a correlation array, signals
from the individual array elements are transmitted to a central location where they are correlated in pairs. This produces interference
fringes which are further processed, principally by the Fourier transform, to produce an image (Thompson, Moran, and Swenson, 1986). A
phased array operates considerably differently; all element outputs
are simultaneously combined together. Phase shifters connected to
each element are adjusted to produce a pencil beam on the sky which is
then electrically scanned over the celestial source in a raster fash-
ion. An image is obtained by displaying the scans on a computer
screen.

Correlation arrays allow considerably larger flexibility in post
processing than do phased arrays. Data can be processed a variety of
ways offline by editing and weighing the correlator outputs differently to form many types of synthesized beams. All data combined from
the many projected baselines during source tracking produce an improved beam because many spatial frequencies are present. With this
increased flexibility, we have more complexity in the receivers, the
array communication system, and the central array electronics.

There are four major drawbacks to the phased array: 1) the beam
shape changes with position, 2) signals from the array elements cannot
be altered in post processing, 3) phase shifters for beam steering are
cumbersome, and 4) interconnection of the elements requires great
lengths of conductors. The interconnecting cables can be eliminated
by transmitting the signals from the elements to the central station
by radio or optics for combination there. However, this requires all
the complications of a correlation array.

We concluded that the correlator array is the most feasible type of
array for the lunar far-side. Receivers manufactured with microelec-
tronic chips utilize very small amounts of power making it feasible to
increase the complexity of the electronics at the elements. Signals
from the elements can be transmitted to the central station which
means that power requirements at the elements will be dominated by the
transmitter.

Array Size

Resolution, and hence the array size, is limited by blurring due to interstellar scintillation. The scattering size of a point source is about one degree at 1 MHz, and about one arcsecond at 30 MHz. Array lengths corresponding to these resolutions are 17 km and 1000 km, respectively. A resolution of one degree at 1 MHz is considerably better than anything presently available.

The array sizes of 17 and 1000 km form natural lower and upper limits for the lunar array. They also conveniently divide the construction of the array and the communications into two phases: Phase 1, in which the inner 17 km of elements are constructed, and Phase 2, in which the elements between 17 km and and 1000 km are constructed. The geometry of the array consists of nonuniformly-spaced array elements whose separations increase with increasing position from the center of the array following a power law. This has proven to be a popular and practical geometry for earth-based arrays. The lunar elements would be spread out two dimensionally over a circular area.

Array Location

A suitable site for the inner 17 km of the array is the crater Tsiolkovsky (Taylor, 1989) with a peak near its center surrounded by a smooth plane 100 km in length. It is located slightly south of the lunar equator near the center of the far-side. From a control station on the peak, signals can be transmitted between the peak and the array elements by line-of-sight vacuum paths. Signals transmitted between the central station and earth must be relayed by lunar satellites.

Antenna elements located in the outer portion of the 1000 km array are beyond the horizon of the central station. Two possible communication paths are relay towers located on the surface, and satellite relays. Towers have the advantage of being stationary, which keeps the phase delays between the elements and the central station fixed.

Due to the long lunar rotation period, data collection for synthesis imaging could take 14 earth days. To keep this time minimal, hundreds of array elements will be deployed. The consequently large number of baselines will assist in reducing celestial interference and confusion since all directivity must come from the synthesis of a pencil beam.

Antenna Elements

Characteristics for ideal array elements include a pattern independent of frequency, no mechanical movement, low sidelobes, high directivity, low mass, and easy deployment. We cannot achieve all of these
simultaneously. High directivity at low frequencies can only be obtained with an element which in itself is an array. The most feasible
compromise is to sacrifice the directivity of the individual elements.

The simplest low-frequency element is an electrical dipole. To
avoid the large power-pattern change with frequency of a near-resonant
dipole (Jasik, 1961), we use an "elemental" dipole 0.1 wavelength
long. Since an elemental dipole has a very fat pattern, we can use a
small phased array of 2 x 2 dipoles as an array element, and phase it
so as to place a null of the 2 x 2 array on an undesirable source.

The radiation resistance of an elemental dipole changes dramatically (a factor of ten thousand) over the operating frequency
range of 1 to 30 MHz. To reduce this change, a 0.1λ dipole at 1 MHz
(30 m long) could be used for the lower frequencies with an electrical
switch along the outer part of each monopole that could be opened to
shorten the dipole for the higher frequencies.

Receivers

The low operating frequency of the telescope permits full technological use of very large scale integrated circuits (VLSI) which reduces mass. A receiver would be placed at each array element with a
self-contained power supply using solar cells in the daytime and
batteries at night.

The receiver design is the usual superheterodyne type. The low
frequency and low interference level on the lunar far-side would permit us to place a 0 to 30 MHz low-pass filter in front of the mixer
and select the observing pass band at the mixer output. Varying the
local oscillator alone would select the observing frequency. Additionally, the local oscillator could be scanned continuously producing a
swept-frequency receiver which is useful for solar observing.

Control signals for the receiver would be received from the central
station via a separate radio communication channel. The synchronizing
local oscillator signal from the central station would be received by

this auxiliary receiver also. The near vacuum on the moon would eliminate any significant phase changes in the propagation path, especially at these low frequencies.

The intermediate frequency (IF) output from each receiver would be transmitted back to the central station. The transmitter for this will be the dominant power-consuming device located at each element. Various alternatives include transmitting at a sufficiently high radio frequency to allow directive antennas to be used, or transmitting with a diode laser coupled to a small optical telescope. A more in-depth study must be made of the alternatives for this communication link which focus on minimum power and minimum hardware at the elements.

Summary

Our preliminary design of a lunar far-side correlation array is based on a two-dimensional nonuniformly spaced array of elemental dipoles extending to a maximum diameter of 1000 km. Phase 1 consists of deploying antennas over a 17 km diameter in the crater Tsiolkovsky on the lunar far-side while the outer antennas are deployed in the second phase. Array operation is over at least four bands from 1 to 30 MHz with variable bandwidths up to 5 MHz and resolutions varying from one degree at 1 MHz to one arcsecond at 30 MHz.

References

Basart, J. and Burns, J. 1989, in A Lunar Far-Side Very Low Frequency Array, eds. J. Burns, N. Duric, S. Johnson, and J. Taylor (NASA Conference Publication #3039), p. 53.

Douglas, J. N., and Smith, H. J. 1985, in Lunar Bases and Space Activities of the 21st Century, ed. W. W. Mendell (Houston: Lunar and Planetary Institute), p. 301.

Jasik, H. 1961, Antenna Engineering Handbook (New York: McGraw-Hill).

Taylor, J. 1989, in A Lunar Far-Side Very Low Frequency Array, eds. J. Burns, N. Duric, S. Johnson, and J. Taylor (NASA Conference Publication #3039), p. 61.

Thompson, A. R., Moran, J. M., and Swenson, G. W. 1986, Interferometry and Synthesis in Radio Astronomy (New York: John Wiley & Sons).

II. INTERFERENCE AT LOW RADIO FREQENCIES

RADIO NOISE NEAR THE EARTH IN THE 1-30 MHz FREQUENCY RANGE

W.C. Erickson[1]

Dept. of Physics and Astronomy

Univ. of Maryland, College Park, MD 20742

Abstract

Terrestrial radio interference presents a serious problem for sensitive low-frequency radio observations from space near the Earth. The interference is both narrow band and broad band. Several satellites and planetary probes have carried radio astronomy experiments so a moderate amount of information is available concerning the noise radiation from the Earth. The region of space within 100 R_E of the Earth is quite a hostile environment for any radio astronomy experiment. Observations up to 10 MHz employing ionospheric shielding may be possible from satellite altitudes on the sunlit side of the Earth near solar maximum. Observations above 10 MHz should be made from the surface of the Earth or from the Moon.

Terrestrial Noise Sources and Emission Levels

Extensive measurements of terrestrial radio noise up to 13.1 MHz were made from Earth orbit aboard RAE-1 (Weber, Alexander, and Stone, 1971) and from lunar orbit aboard RAE-2 (Alexander et al., 1975). LaBelle, Treumann, Boehm, and Gewecke (1989) discuss more recent measurements up to about 5 MHz made from various spacecraft.

Spectacular evidence concerning the severity of terrestrial interference is shown in Figure 1. It is a "typical" record from RAE-2 as the Earth is occulted by the Moon. One sees that even from the distance of the Moon, the radio noise in the 1 to 10 MHz range is dominated by

---

[1] Present address:

Department of Physics
University of Tasmania
GPO Box 252C
Hobart, Tasmania 7001
Australia

terrestrial interference.   The interference levels that would be encountered by a satellite 10,000 km above the Earth's surface (24 times closer to Earth than the Moon) would be some 28 db higher than those shown in Figure 1.

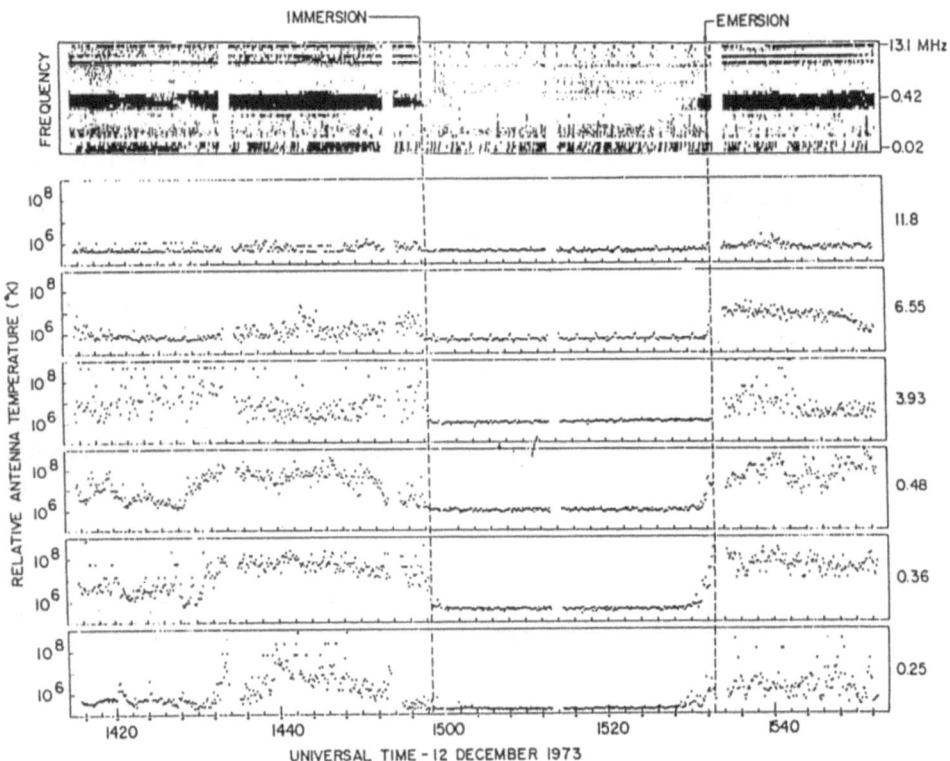

Figure 1.   RAE-2 data taken from lunar orbit.   At all of the frequencies observed the noise levels are dominated by terrestrial noise except when the Earth is occulted by the Moon's limb.   When the Earth is occulted, the noise levels are dominated by the Galactic background.   Immersion and emersion are calculated for the center of the Earth.   Some of the terrestrial noise sources, such as auroral kilometric radiation (AKR) below 1 MHz, are located 1 to 2 R_e from the Earth on the nighttime side and radiate past the calculated position of the limb. (from Alexander et al, 1975)

Radiometer noise levels are normally dominated by the Galactic background and this background level is used as a reference noise level. Near the Galactic Poles the background has a brightness of about 10^{-20} w m^{-2} Hz^{-1} ster^{-1} in the 2 to 10 MHz range.   The brightness temperature varies approximately as (wavelength)^2, from about 2×10^7 K at 1 MHz to 2×10^5 K at 10 MHz.   Near the Galactic Plane the background temperatures are lower due to free-free absorption, but the above values would be typical antenna temperatures for low-gain dipolar antennas.

Herman, Caruso, and Stone (1973) made a specific study of terrestrial radio noise in the 4 to 10 MHz range with RAE-1. Their study employed the downward-pointing "V" antenna which had a typical beamwidth of ≈30° and their data refer to the radio noise emanating from the subsatellite region of the Earth. They found that terrestrial noise levels were lowest when RAE-1 was over the oceans and the highest intensities were recorded over major northern and southern land masses. From the noise characteristics they surmised that over the United States the principal source is man-made noise from populated areas (ignition noise, electrical machinery, etc.), while over South America it is lightening. Ground-based transmitters apparently dominated the noise levels over Asia and Eastern Europe.

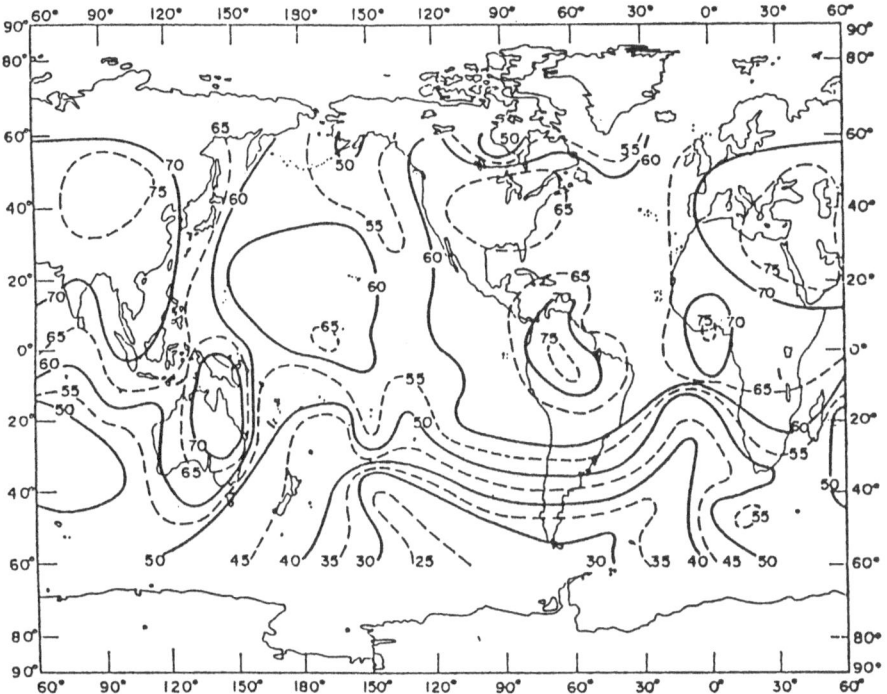

Figure 2. The terrestrial radio noise distribution derived from the RAE-1 (height 6000 km) lower "V" data at 9.18 MHz for December 2-6, 1968. The secondary peaks in activity over the mid-Pacific and northern Australia are believed to be correlated with local thunderstorm activity. Contour levels are db above 288 K. The Galactic background on this scale would be about 31 db and the receiver saturated at 75 db. (from Herman et al, 1973)

A typical global noise distribution at 9.18 MHz is shown in Figure 2. It appears that natural and man-made noise emissions are more-or-less comparable in level. This means that operation of a satellite system in an allocated radio-quiet band, even if effective policing of the band were possible, will not solve the interference problems because much of the noise is broad band static from unlicensed sources. Antenna temperatures observed by RAE-1 at nighttime over central United States agree well with Horner's (1965) estimates of the interference levels to be expected from lightening; these levels are about 35 db above the Galactic background. On control nights (without thunderstorms) the observed levels were about 25 db above background and agreed with those predicted for man-made noise levels from electrical machinery.

LaBelle et al. (1989) have made a more recent study employing the AMPTE/IRM spacecraft in the 1.0-5.6 MHz range at a distance of 10 to 18 R_z. They suggest that terrestrial radiation levels may now be approximately 20 db higher than those measured by Herman et al. in 1968.

Terrestrial interference has prevented scientific investigations of natural sources from satellite altitudes in the 10-30 MHz range but there is not much information available about these interference levels. Spacecraft receivers in this band are often saturated until the spacecraft are well away from the Earth. However, most of the sources of terrestrial noise below 10 MHz, such as transmitters, electrical machinery, and lightening, are not significantly weaker in the 10-30 MHz range. Cosmic sources, having non-thermal spectra, are considerably weaker relative to the interference.

Another natural source of intense low frequency noise is auroral kilometric radiation (AKR) (Gurnett, 1974; Kaiser and Alexander, 1977; Benson, 1985). This radiation can be 90 to 100 db above the Galactic background. It occurs primarily below 1 MHz on the nighttime side of the Earth. First harmonic emissions of AKR were observed up to 1 MHz by LaBelle et al (1989) but they do not find evidence for higher harmonics. Therefore, it appears that AKR is not a particularly grave problem above 1 MHz.

It should be noted that ISEE-3 carried a radio receiving system that operated in the 30 kHz to 2 MHz band. This satellite was located at the inner libration point, ≈240 R_z from Earth towards the Sun where it was shielded by the ionosphere. It made excellent solar observations for several years without any particular problems caused by terrestrial interference. Although ISEE-3 was much less sensitive than any proposed low frequency space array, its highly successful operation suggests that sensitive observations up to a few MHz should be possible in regions of space shielded by the ionosphere.

63

Ionospheric Shielding

The ionosphere should effectively block all ground-level emissions at frequencies below f_oF_2. Electromagnetic waves below this ionospheric plasma frequency become evanescent and die away with skin depths of only tens of meters. The ionosphere should become extremely opaque and attenuate such waves by thousands of decibels.

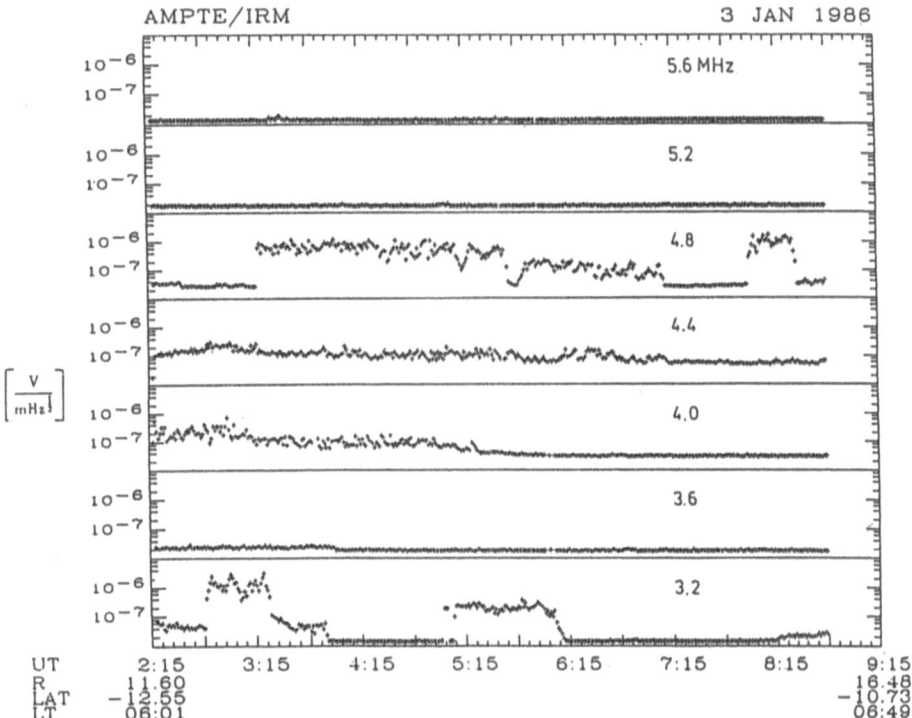

Figure 3. AMPTE/IRM data showing two-minute averages of the rms spectral intensity versus time for seven frequency bins (3.2-5.6 MHz), each 10 kHz broad, for 0215-0915 UT on January 3, 1986, a typical time interval away from the noon local sector. The satellite was at distances from 11.60 to 16.48 R_E. Values of rms spectral intensity 10-30 db above the noise level occur in bursts lasting from a few minutes to a few hours. The absolute intensity of these bursts is in general higher at distances nearer to the Earth. (from LaBelle et al, 1989)

Typical global distributions of f_oF_2 suggest that ionospheric shielding should be effective up to at least 10 MHz on the sunlit side of the Earth near solar maximum and up to at least 3 MHz near solar minimum. At frequencies somewhat above f_oF_2 the ionosphere strongly refracts and reflects the waves, usually allowing them to reach a

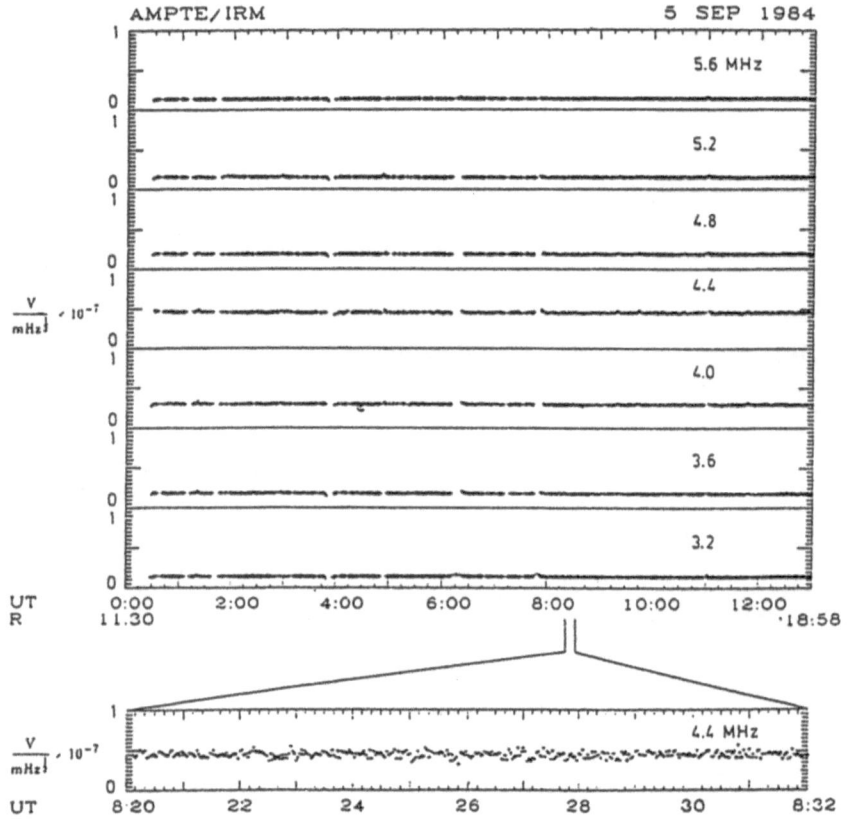

Figure 4. AMPTE/IRM data showing two-minute averages of the rms spectral intensity for seven frequency bins (3.2-5.6 MHz) on a linear scale for a 12-hour interval on September 5, 1984, when the satellite was near noon local time and moved outwards from 11.3 to 18.6 R_E. The bottom panel presents an expanded view showing 2-second resolution data (at 4.4 MHz only) from a 12-minute interval on the same day. The noise level observed near noon local times is constant except for fluctuations of about 10% at time scales of a few seconds and much less over longer time scales. No radial dependence of the noise level is observed. (from LaBelle et al, 1989)

satellite only if the source is located near the sub-satellite point. However, terrestrial radiation escaping through the ionosphere at some distant point can sometimes be ducted to the vicinity of a satellite that would otherwise be shielded. The RAE-1 and AMPTE/IRM data both show cases where terrestrial noise broke through the ionosphere at frequencies where it would not be predicted. Nevertheless, both these data sets also show long periods when the satellite was shielded and the Galactic background level was recorded. Figures 3 and 4 illustrate these situations.

Solar and Jovian Emission

Ionospheric shielding will be effective only on the sunlit side of the earth, where solar emission may interfere with observations. Type III solar bursts are the most common. These bursts have average durations at ≈1 MHz of about ten minutes; at ≈10 MHz their durations are a few minutes or less. They occur at a typical rate of several per hour at solar maximum and at average intervals of many hours at solar minimum. Weak bursts are a few decibels above the Galactic background on low gain antennas; strong bursts are 30-60 db above background.

Jovian bursts can be intense but they are short-lived, they occur during predictable periods, and come from a definite direction. They may be a nuisance but it should be possible to cope with any problems that they present.

I have examined the statistics of these bursts using ISEE-3 data spanning a period of solar maximum from launch and positioning at the inner libration point, L1, in September, 1978, until the spacecraft left L1 at the end of 1982. The spinning dipoles of the spacecraft provided a rudimentary direction-finding capability which indicated that almost all of the observed bursts came from the direction of the sun. At the three highest ISEE-3 frequencies, the percentages of the time when the radiometer outputs exceeded the Galactic background by 3, 6, and 10 db are given in Table 1.

TABLE 1

Frequency	1980 kHz	1000 kHz	513 kHz
3 db above background	10.4±0.8%	20.7±1.4%	43.1±2.2%
6 db above background	3.9±0.4%	9.5±0.9%	26.9±1.9%
10 db above background	1.6±0.2%	4.0±0.4%	13.7±1.2%

These data indicate, for example, that if a system were to operate at 1980 kHz, and all data were to be discarded when the level of solar emission exceeded the Galactic background by 3 db, it would be necessary to discard only about 10% of the data. Solar bursts have shorter durations as the frequency increases, so observations at frequencies

above 1980 kHz should suffer from less data loss due to solar interference. It would be necessary to develop interference excision
routines to remove the effects of solar emission near or below the
Galactic background level.

Interference Levels Harmful
to a Low Frequency Interferometer in Space

The CCIR definitions of harmful interference have proven to be
highly appropriate for radio astronomy observations. The CCIR
definition of the interference level which is harmful to radio astronomy
observations is 10% of the RMS noise limit of a simple, filled-aperture
radio telescope. The harmful interference levels for interferometers
were analyzed by Thompson (1982) who calculated the interference level
that would add 10% to the noise fluctuations in a map produced by a
synthesis telescope. Thompson also conducted an experiment with an
artificial interfering source located on a mountain overlooking the VLA
in order to verify his calculations. He found that the VLA at 1400 MHz,
in its most compact configuration, is about 14 db less sensitive to
interference than a filled-aperture system. In its most extended
configuration, it is about 22 db less sensitive. This results primarily
from the reduction of the system's response to an interference source
at zero fringe frequency when the system is introducing fringe rotation
to follow the sidereal motion of an astronomical source. The interfering signal is rotated in phase over many cycles and its effects are
greatly diminished. The reduction factor for each interferometer
baseline is given by

$$F_i = \mathrm{sinc}(\tau f_i)$$

where f_i is the fringe frequency on the i'th baseline and τ is the
averaging period for the data on this baseline. The total reduction
factor is found by averaging F_i over all baselines.

The data are gridded in the u,v plane before Fourier transformation
and τ is determined by the length of time required for the i'th baseline
to rotate across each cell in this grid. The grid size is equal to the
reciprocal width of the synthesized field. For large x, $\mathrm{sinc}(x) \approx 1/x$
so F_i is small if (τf_i) is large. The interference perturbs most
strongly those interferometer baselines having small (τf_i). In the case
of the VLA this product is >>100 for most baselines. The contributions
from those baselines having low fringe rates results in "noise" which
is mostly in the form of low frequency ripples across the synthesized
map.

If low gain dipolar antennas are used in a space array it will be necessary to synthesize about 2π steradians of the sky in order to deconvolve the sidelobe effects of numerous strong sources within the primary patterns of the antennas. This means that very small cell sizes are required. The data must be gridded into cells no more that one wavelength across if a 180° field of view is to be synthesized. Thus, only a few radians of fringe phase would occur within each integration period. (τf_i) will be ≈1 and fringe rotation will not be very effective in reducing the system's response to harmful interference.

It may be possible to use coherence effects and fringe-rate smearing to reduce the effective field-of-view to a steradian or less. This would allow a somewhat larger cell size and somewhat longer integrations. I estimate that the harmful interference limit will be only 3 to 10 db above that of a filled-aperture system. A more exact estimation would be difficult without a detailed system design. (Note that coherence effects will not reduce the system's responses to narrow band signals; these responses can only be reduced by fringe rotation.)

VLBI experience with interference is not applicable to a low frequency interferometer in space. VLBI systems are especially insensitive to interference because the interfering signal does not correlate at the two ends of the interferometer. Unfortunately, terrestrial interference will be fully correlated on all interferometer baselines of a space array. Thus, a low frequency interferometer in space will be highly susceptible to harmful interference.

In order to be really useful, the low frequency array must be capable of observing sources at the 1 to 10 Jy level and the harmful interference level would be 0.1 to 1 Jy. For example, if one assumes a harmful interference limit of 0.5 Jy and a frequency of 3 MHz, then on a single satellite's dipole this signal will produce an antenna temperature of 0.23 K and if the array has 10 satellites this gives an equivalent filled-aperture antenna temperature of 2.3 K. Assuming that the interferometer provides 10 db protection compared to a filled aperture, the harmful interference limit would correspond to an antenna temperature of 23 K. This is some 50 db below the Galactic background temperature. All of the previous estimates of interference levels were a few decibels to 70 db above the Galactic background. Thus, shielding or interference rejection at the level of 50 to 120 db would be required for successful operation. The development of interference rejection techniques to this level is not feasible so ionospheric shielding must be employed. Any small leakage of interference through the ionosphere would destroy the data unless sophisticated interference rejection techniques are used as well.

Conclusions

In the 1 to 10 MHz range ionospheric effects make high-angular resolution (≈arc-minute) observations from the Earth's surface virtually impossible and it is necessary to go to space. At satellite altitudes near the Earth, terrestrial sources generate noise levels millions of times above the level of interference harmful to observations with a low-frequency space array. This interference is both narrow band and broad band noise. It is essentially impossible to filter or excise interference to such levels.

Near solar minimum, sensitive observations up to a few MHz may be possible from Earth orbit on the sunlit side of the Earth where the ionosphere should provide sufficient shielding from terrestrial interference. Near solar maximum it may be possible to work up to 10 MHz from the sunlit side, at least during selected periods.

In the 10 to 30 MHz range sensitive observations from space near the Earth are probably impossible because ionospheric shielding will be ineffective. In this case, Earth-based observations utilizing terrain shielding are far more practical. The problems associated with ionospheric refraction in Earth-based observations can be attacked with modern self-calibration techniques that have proven to be highly effective at higher frequencies. This approach is far more feasible than an attempt to cope with the interference levels at satellite altitudes.

The interference levels at the near side of the Moon will be about a thousand times lower than those at typical satellite altitudes. This reduction in levels may make relatively sensitive observations possible if effective interference rejection techniques are developed.

The far side of the Moon appears to be the only location near the Earth that is sufficiently shielded from terrestrial interference to permit observations without interference rejection systems. In the distant future, when the severe communication and logistical problems associated with the lunar far side are solved, it is the most promising site for low-frequency radio astronomy.

Acknowledgements

This paper is based on a report prepared for the Jet Propulsion Laboratory (JPL Publication 88-30). I wish to acknowledge very helpful discussions with Mike Kaiser, Bob MacDowall, and Bob Stone at GSFC, with Dayton Jones, Tom Kuiper, Mike Mahoney, and Mike Janssen at JPL and with Hilary Cane. However, this is not to dilute my responsibility for all errors and omissions. The ISEE-3 data were provided to me by Bob Stone and Bob MacDowall of the Radio Astronomy Group at GSFC.

References

Alexander, J.K., Novako, J.C., Grena, F.R., and Weber, R.R., 1975, "Scientific instrumentation of the RAE-2", Astron. and Astrophys. 40, 365.

Benson, R.F., 1985, "Auroral kilometric radiation: Wave modes, harmonics, and source region electron densities", J. Geophys. Res. 90, 2753.

Gurnett, D.A., 1974, "The Earth as a radio source; Terrestrial kilometric radiation", J. Geophys. Res. 79, 4227.

Herman, J.R., Caruso, J.A. and Stone, R.G., 1973, "Radio Astronomy Explorer (RAE)-I. Observations of terrestrial radio noise", Planet. Space Sci. 21, 443.

Horner, F., 1965, "Radio noise in space originating in natural terrestrial sources", Planet. Space Sci. 13, 1137.

Kaiser, M.L. and Alexander, J.K., 1977, "Terrestrial kilometric radiation, 3, Average spectral properties", J. Geophys. Res. 82, 3273.

LaBelle, J., Treumann, R.A., Boehm, M.H., and Gewecke, K., 1989, "Natural and Man-made Emissions at 1.0-5.6 MHz Measured Between 10 and 18 R_z", MPE Preprint 146 (accepted by Radio Science).

Thompson, A.R., 1982, "The response of a radio-astronomy synthesis array to interfering signals", IEEE Trans. Antennas Propag. AP-30, 450.

Weber, R.R., Alexander, J.K., and Stone, R.G., 1971, "The radio astronomy explorer satellite; A low frequency observatory", Radio Sci. 6, 1085.

A QUANTITATIVE ASSESSMENT OF RFI IN THE
NEAR-EARTH ENVIRONMENT:

Michael D. Desch
NASA Goddard Space Flight Center
Greenbelt, MD 20771

Central to the issue of making observations from space near the earth is the assessment of the environmental noise levels at very low frequencies. It is well known from previous space-borne observations, particularly from the Radio Astronomy Explorer satellites, that the radio-frequency interference (RFI) levels below about 10 MHz are very high, sometimes exceeding the galactic background noise by at least 40 dB. In this paper I will review the known sources of rf noise, their spectral and intensity characteristics, and how serious a threat each represents to monitoring in the earth's neighborhood. It is clear that the quietest frequency/time regime for monitoring galactic and extragalactic sources is over the dayside of the earth in the 0.8 – 10 MHz band. Observations made in 1969 from RAE-1 achieved an interference-free duty cycle of nearly 50% in the band below 4 MHz on the nightside of the planet. It is felt that by going to an altitude of ~ 10R_e, observations during the present epoch could match or exceed the 1969 RAE observing conditions, particularly on the dayside of the planet.

Introduction

In the near-earth environment, there are three major sources of noise that are capable of disrupting observations of weak radio sources at very low frequencies. These are spherics, solar type III bursts, and the earth's auroral kilometric radiation (AKR). Spherics, under which heading I include both natural and man-made sources, originate on or near the surface of the earth. Solar type III bursts are generated in and above the outer solar atmosphere. The AKR is generated at high magnetic latitudes on the earth's nightside. Less important sources of noise include antenna thermal noise, plasma waves, and radio emissions from the known planetary sources: Jupiter, Saturn, Uranus and Neptune.

A proper assessment of the ability to adequately perform deep-sky interferometric monitoring of radio sources at low frequencies near the earth requires that this hostile environment be examined rather closely. One previous study (Erickson, 1988) came to the conclusion that the noise problem is extremely severe and may constitute an inescapable impediment to deep-sky observations. In this study, we examine the spectral and temporal characteristics of the major noise sources to attempt to identify those frequency bands and observing times that avoid some or all of the interference. From a re-examination of RAE satellite data (see Weber *et al.*, 1971 for an instrument description), it is found that sufficiently quiet conditions existed on the *night* side of the planet in 1969 to detect Jupiter emission on a fairly regular basis. We conclude that contemporary observations could easily match the earlier RAE observations if care is taken to confine the monitoring in frequency and local time, and if pre-amps are selective enough to reject large signals outside the desirable bands.

Sources of RFI

Figure 1 shows an overview of the principal sources of noise in the terrestrial environment. Each spectrum is a measure of the flux density as would be observed by an instrument in the vicinity of the earth. In the case of the AKR, the distance to the source is normalized to $25R_e$. In the following section, each noise source will be discussed in the context of possible deep-sky observations from space. Table 1 at the end of this section summarizes the principal characteristics of each.

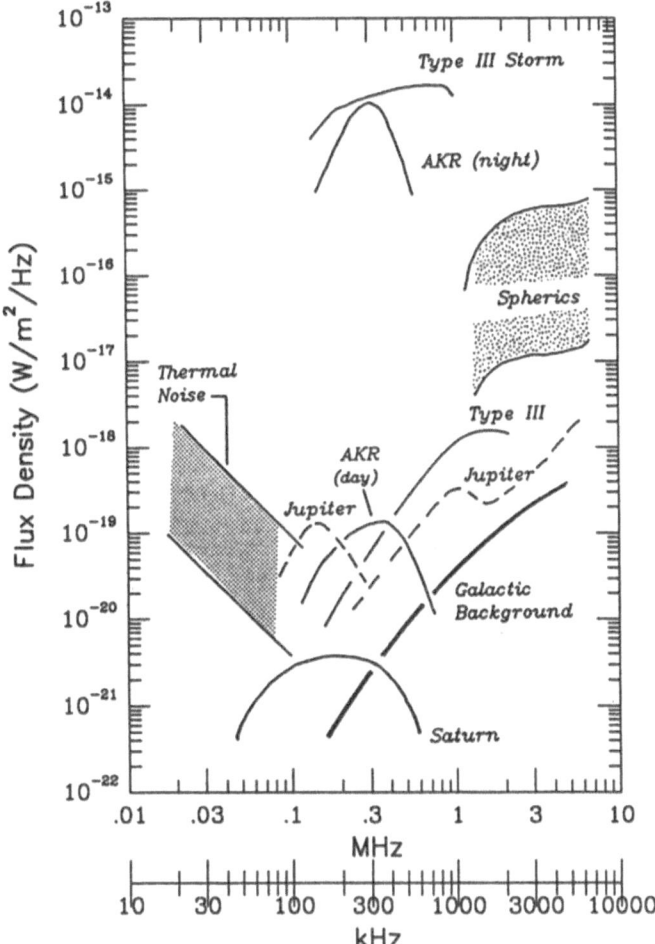

Fig. 1. Overview flux spectra of the principal sources of noise in the terrestrial environment below 10 MHz. The flux density of each source is normalized such that it shows the intensity of each source as it would appear from earth. In the case of the AKR, the distance is normalized to a source-spacecraft distance of 25 R_e.

Spherics. The most serious threat to space-borne monitoring at low frequencies comes from the rf noise generated by man-made and lightning emissions. I include both man-made and lightning noise in the category of spherics because both are generated in the atmosphere, in contrast to all the other sources of interference which are *magneto*-spheric in origin, either terrestrial, solar, planetary, or local. The man-made noise sources consist of machinery, ignition discharges, sweeping radars and, principally, communications transmitters. Natural spherics are due to the combined effects of lightning from world-wide thunderstorm activity. The latter is most severe in the equatorial regions of the earth and in the evening local time sector.

Global measurements of the spherics noise level at 9.18 MHz were made by Herman *et al.* (1973) using the lower V antenna on the RAE-1 satellite. Their map of the noise contours is shown in Figure 2. The galactic background equivalent corresponds approximately to 30 dB in this figure. The 200–kHz bandwidth receivers saturated at 75 dB. It is clear that at some times the peak noise levels exceeded 45 dB above the galactic background. As seen by RAE-1, the spherics noise level was most severe at 9 MHz because below this frequency the ionosphere begins to provide significant shielding, particularly on the dayside of the planet.

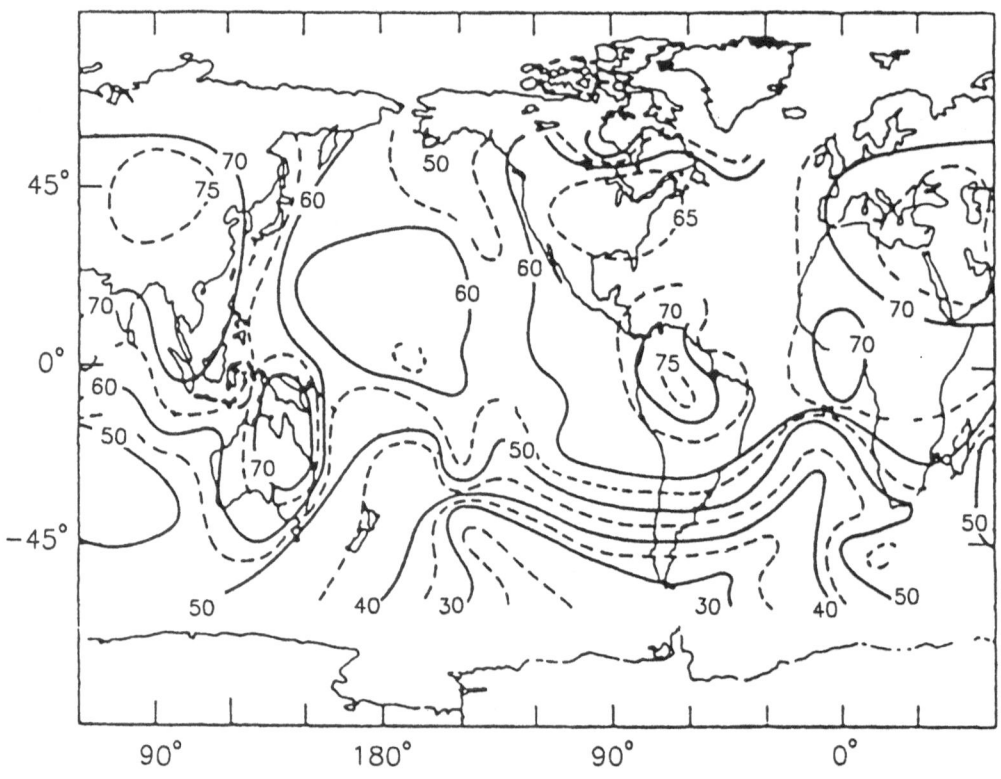

Fig. 2.   Global survey at 9.18 MHz of man-made and lightning noise from the lower-V antenna on RAE-1 . (Figure adapted from Herman *et al.*,1973.)

Solar Bursts. Solar type III bursts are generated by streams of high energy electrons as they sweep outward from the sun. The characteristic negative frequency drift exhibited by type III bursts is due to the decreasing plasma frequency encountered by the stream as it propagates outward. Type IIIs are the most intense of the natural emissions observed near the earth and during solar storms can represent a nearly continuous background of enhanced radio emission. They are a more serious problem below about 1 MHz, where the frequency drift tends to smear the emission in time (see figure 3). At higher frequencies the signal is more short lived. During solar quiet times, which are more common during solar minimum, type III bursts are likely to be unobserved for long intervals, even at the lowest frequencies. Type IIIs have a smooth temporal signature, characterized by a relatively sharp onset and slow decline.

AKR. The earth's auroral kilometric radiation (AKR) is predominantly a nightside, high magnetic latitude phenomenon. Although the total amount of energy available to the radiation process is small compared to the solar emission processes, the generation of AKR is relatively efficient since it is believed to be due to a maser operating at the electron cyclotron frequency. Since it is generated on the nightside of the earth and is beamed predominantly outward, the AKR is seen almost exclusively by spacecraft over the planet's nightside. Additionally, the AKR is confined to frequencies below ~750 kHz (see figure 3) except for occasional exotic emission conditions. Therefore the local time and frequency band limitations of AKR help to minimize the impact of this natural component on deep-sky observations. AKR has a very bursty emission character and is elliptically polarized.

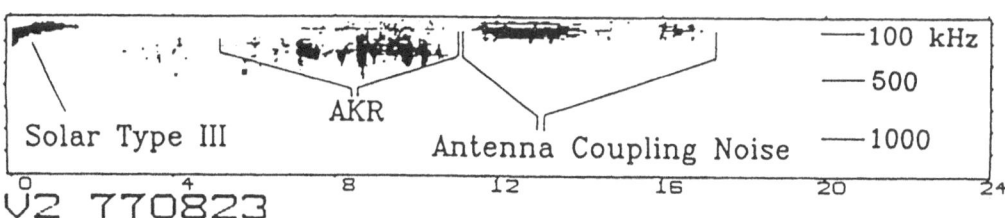

Fig. 3. Radio spectrogram from the Voyager spacecraft showing an example of a type III burst, AKR, and antenna coupling noise (thermal electron noise). Only the tail end of the type III burst is visible.

Electron and Plasma Noise. At the lowest frequencies at which propagating emissions can be observed, radiations due to the presence of the antenna in a plasma and plasma waves generated by solar wind shocks are seen. The noise induced by the presence of an antenna in a plasma is commonly referred to as antenna thermal noise. Under certain conditions, the thermal motion of charges particles induces a variable electrostatic potential on the antenna. In effect, the antenna becomes an electron spectrometer that measures the local electron properties. Antenna thermal noise is most commonly seen during times when the solar wind density at 1 AU is rather high, that is when the plasma density exceeds, say, 100 cm^{-3}, corresponding to plasma frequencies near 100 kHz (see figure 3). In the earth's vicinity, antenna thermal noise is not usually observed above ~200 kHz.

The plasma noise referred to here is that often observed in association with shocked solar wind. These are intense narrow-band electrostatic emissions near the local plasma frequency. They are sometimes seen to accompany solar type III bursts and are often of very brief duration. These waves are thought to be a consequence of plasma turbulence effects induced by a local instability. The frequency range is similar to that of the antenna thermal noise.

TABLE 1

Low Frequency Noise Sources at 1 AU

Source	Frequency	Polarized	Intensity	Spacecraft
Spherics	1 MHz - daylite	Yes	V Strong	all
Type IIIs	20 kHz-daylite	No	V Strong	all
AKR	20-750 kHz	Yes	V Strong	all
Electron & Plasma	dc-200 kHz	No	Moderate	all
Jupiter	20 kHz-40 MHz	Yes	Moderate	V1,V2,ISEE, RAE
Saturn	20 kHz-1 MHz	Yes	Moderate	V1, V2, ISEE
Uranus	20 kHz-850 kHz	Yes	Weak	V2
Neptune	20 kHz-2 MHz	Yes	Weak	V2

Planetary Emissions. Jupiter has been known to be an intense source of radio emission at decameter wavelengths since 1955. Spaceborne observations have shown that Jupiter, as well as the other major planets, Saturn, Uranus, and Neptune, are important sources of emission at kilometer and hectometer wavelengths as well. None of the planets other than Jupiter could be considered important *interference* sources, however, and even Jupiter, while it is sometimes as intense as the sun, is only observed very infrequently compared with the terrestrial and solar radiations. In fact, Uranus and Neptune would be extremely difficult to detect from earth orbit because, not only are they so much farther away, but they are intrinsically weaker as well. Saturn, on the other hand, is observable and has been detected by ISEE-3 at the earth-sun libration point. Therefore, while the outer planets do not constitute possible sources of interference, Jupiter and Saturn are possible targets for deep-sky monitoring

by interferometric means. The Saturn emission is confined to frequencies below ~1 MHz, whereas the Jovian emission is observable over a broad range of
frequencies from 40 MHz down to the local solar wind plasma frequency.

Interference-Free Monitoring

To estimate the percentage of interference-free monitoring time that one
might expect at low frequencies, RAE-1 observing statistics from 1969 have
been re-analyzed. Based on an analysis by Desch and Carr (1978) who were
searching for evidence of Jovian emission during a 128–day interval from 13
Jan — 21 May, 1969, 8 RAE-1 frequency channels have been re-examined
with a view toward establishing observing statistics below 10 MHz. The Ryle-
Vonberg receivers, which have 40 kHz bandwidth and 0.1 sec time constant,
were used throughout the analysis. A requirement in the original study was
that the observing conditions be extremely quiet in order to reliably identify the
Jovian noise in the presence of the rather hostile rf background discussed so far.
The satellite observing geometry was optimized to minimize the earth-Jupiter
distance and to minimize interference from the sun (type III bursts), therefore
the observations were centered on the time of Jupiter opposition. That is, from
the standpoint of spherics and AKR, the observations were less than optimal
because the spacecraft was over the nightside of the earth when the RAE upper-
V antenna was sweeping across Jupiter.

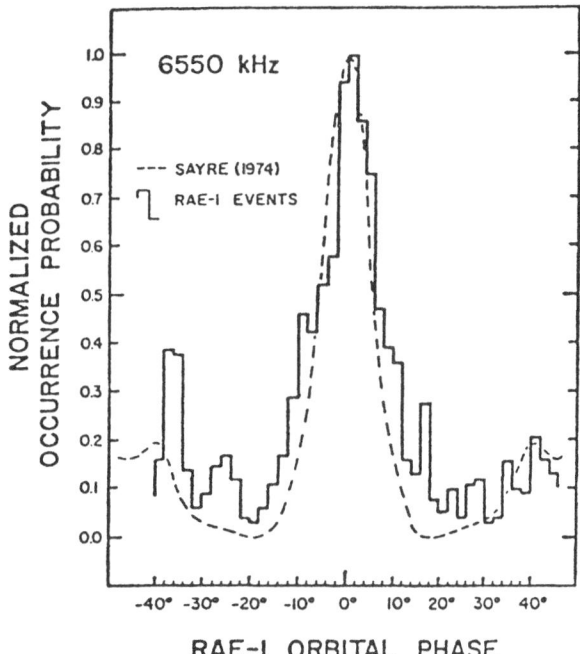

Fig. 4.   Normalized detection probability of Jupiter events is compared with the
calculated V-antenna pattern.

Despite this, a measure of the success with which the Jupiter bursts were identified is apparent from Figure 4. Here, the candidate Jupiter bursts were stacked as a function of the satellite orbital phase. In this figure, 0° phase corresponds to the time when the V antenna beam is aligned with Jupiter. The number of counts (solid line) is compared with the theoretical V-antenna beam pattern as computed by Sayre (1974). It is clear that the variation in the number of detected Jupiter events as a function of orbital phase closely matches the antenna pattern itself, as one would expect if the identifications are correct. This result would not have been possible if the background noise were overwhelming. In fact, there was a considerable volume of interference-free observing time. In Table 2 we show what volume of interference-free monitoring was obtained:

Table 2

Low-Frequency Observing Statistics

Frequency (kHz)	Total Observing (hr)	Interference Free (hr)	%	Limitation
6550	355	111	31	spherics
4700	590	206	35	spherics
3930	590	276	47	spherics
2200	1520	712	47	spherics
1310	1520	684	45	spherics
900	1520	389	26	AKR
700	1520	285	19	AKR
450	1520	63	04	AKR

In the 1–4 MHz band, interference-free monitoring reached nearly 50%, while reasonably good conditions extended up to over 6 MHz (30–35%). Above 6.55 MHz, that is at 9.18 MHz (the next RAE channel) ionospheric breakthrough was a very serious problem so that no reliable observations were possible. Of course, these observations pertained to nighttime observing conditions, as mentioned, so this is not surprising. Below about 1 MHz, conditions deteriorated rapidly due to the presence of AKR.

No statistics exist as yet for dayside observing; however, the presence of spherics would be far less problematic due to increased ionospheric shielding. Therefore, observing in the 1–9 MHz band would be considerably improved. Even on the dayside, however, ionospheric breakthrough would be a nearly contin-uous source of interference above 10 MHz, limiting observations to only the strongest sources: type III bursts and possibly Io-stimulated Jupiter emissions, for example. An example of what dayside monitoring at altitudes much greater than 1 R_e might be like is shown in Figure 5. This figure shows a radio spectro-gram of data from the IMP-6 satellite in earth orbit. Jupiter emission is clearly visible in the band between 1 and 10 MHz which is generally clear of interfer-ence from spheric and type III bursts.

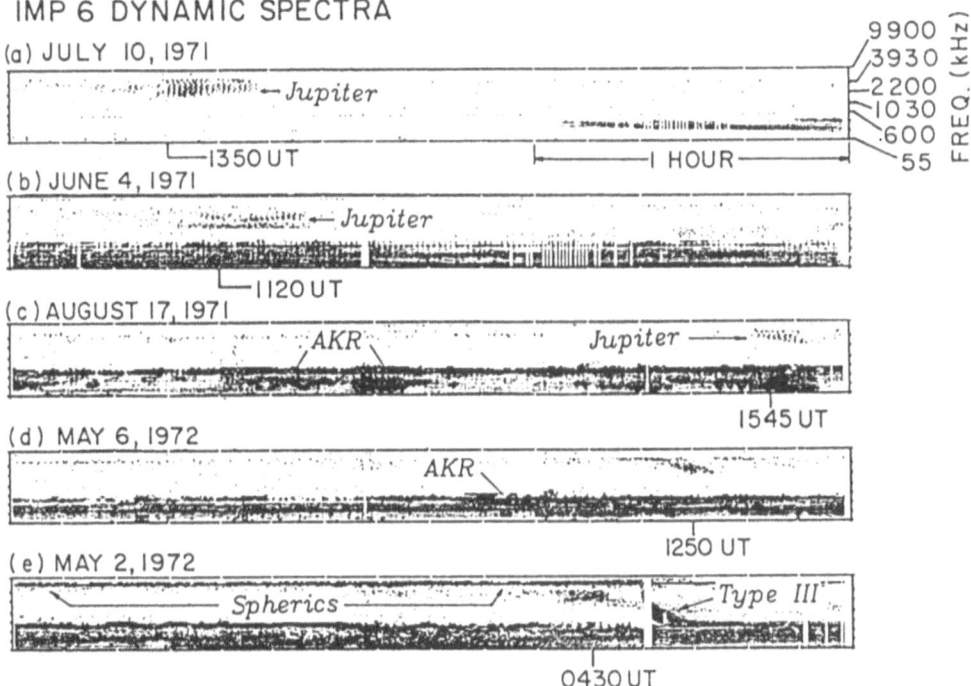

Fig. 5. Radio spectrogram of data from the IMP-6 satellite showing Jupiter emission as well as AKR, a type III burst and spherics.

Summary

It is clear that, as mentioned earlier, the major sources of interference are those due spherics (natural and man-made), solar type III bursts, and AKR. Because the type III bursts are only a moderate problem above 1 MHz and then only a portion of the time (during active solar periods), and because the AKR is confined spatially to the nightside of the earth and in frequency space to < 750 kHz, radiation from spherics poses the most serious threat to near earth observations. The only relief from spherics is provided by the limited shielding capability provided by the earth's ionosphere. Experience with RAE-1 shows that over the nightside of the earth, the ionosphere becomes relatively transparent, as expected, so that spherics are observed down as low as 1–2 MHz. However, moderately good observing conditions can be found, even over the nightside, at frequencies between 1 and 5 MHz. Over the dayside of the earth, ionospheric shielding is much more effective in interrupting the propagation of signals from the ground, so that quiet observing conditions should be expected >50% of the time at frequencies up to at least 6 MHz or so. Above this frequency the ionosphere becomes somewhat more 'leaky'. At frequencies near 10 MHz, one should not expect quiet conditions more than about 10 — 15% of the time.

These conclusions are based on observations made during the 1969 apparition of Jupiter at a satellite altitude of 1R_e. Since this time, conditions with respect to spherics have certainly been aggravated by the continuing onslaught of human development and consequent crowding of the available communications spectrum. Therefore, strictly speaking, the observing conditions at 1 R_e have probably deteriorated relative to the numbers shown in Table 2. On the other hand, because of other considerations, namely birefringent propagation of incoming waves through the ambient plasma, deep-sky interferometric observations would not be made at this low an altitude. At an altitude of, say, 10 R_e, the spheric noise level would be reduced by 20 dB relative to that at 1 R_e, possibly more than making up for the increase in man-made interference since the 1969 epoch. And, of course, observations from the lunar vicinity would be cleaner still.

References

Desch, M. D. and Carr, T. D.: 1978, *Astron. J.* **83**, 828.
Erickson, W. C.: 1988, JPL Publication 88–30.
Herman, J. R., Caruso, J. A. and Stone, R. G.: 1973, *Planet. Space Sci.* **21**, 443.
Sayre, E. P.: 1974, AVCO-Systems Report, AVSD—0144–74–CR.
Weber, R. R., Alexander, J. K. and Stone, R. G.: 1971, *Radio Sci.* **6**, 1085.

STATUS OF LOW FREQUENCY RADIO ASTRONOMICAL

ALLOCATIONS AT THE 1992 WORLD ADMINISTRATIVE

RADIO CONFERENCE (WARC-92)

Tomas E. Gergely

National Science Foundation, 1800 G Street, N.W.

Washington, D.C., U.S.A.

Pressures on the radio spectrum increased enormously since the last World
Administrative Radio Conference (WARC), which dealt with frequency allocations over
the entire radio spectrum, was held in 1979 (the International Telecommunication
Union [ITU] allocates the spectrum in the 9-kHz to 275-GHz range). Since WARC-79,
several conferences were held to deal with allocations in selected portions of the
spectrum or with special issues such as the geostationary orbit. The spectrum
needs of many emerging new technologies and services were left unsatisfied by
these conferences. The Plenipotentiary Conference of the ITU, held in June 1989,
decided to convene another WARC, to be held in Spain in the first quarter of 1990,
in order to resolve these issues. The Agenda of WARC-92 will take into account the
Resolutions and Recommendations of some previous Conferences: those of the High
Frequency Broadcast Conference held in 1987 (HFBC-87), the WARC for the Mobile
Services held in 1987 (WARC MOB-87), and the WARC for the Geostationary Orbit
held in 1988 (WARC Orb-88). A definitive Agenda for WARC-92 will be established
by the meeting of the Administrative Council of the ITU, to be held in June 1990.

The high-frequency (HF) band, defined as 3 to 30 MHz by the Radio
Regulations (RR), is one of the most congested regions of the spectrum. Numerous
radio services; e.g., the aeronautical and maritime mobile, fixed, broadcasting,
and amateur services operate in the band, to mention only a few. In the U.S.,
the number of Government frequency assignments in the HF band increased by
approximately 50 percent during the period 1980-1987; the trend of private-sector
assignments is possibly similar. (In the U.S., Government- and private-sector
assignments are recorded separately.) The trend is illustrated in Fig. 1, which
shows the increase in frequency assignments recorded in the Government Master File
in those HF subbands shared by the fixed and mobile services on a coequal, primary
basis [1].

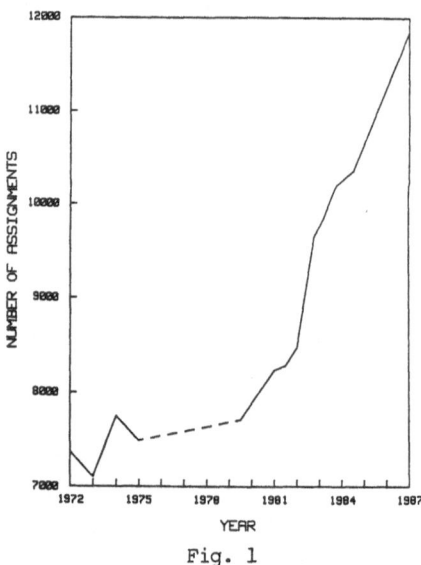

Fig. 1

Besides the increased activity of several traditional users of the band, such as the fixed and mobile services or the broadcasting service, a number of new technologies have also come into existence. Over-the-horizon radar and meteor-burst systems are two examples of such technologies. While these technologies are not entirely new, their use became practical thanks to the tremendous increase in computing capability and simultaneous drop in price experienced during the last decade.

In the HF region of the spectrum, two bands were allocated to radio astronomy at WARC-79:

$$13,360-13,410 \text{ kHz}$$

and

$$25,560-25,670 \text{ kHz}.$$

In the U.S., both allocations are primary, exclusive. Internationally, the lower band is shared with the fixed service in all three regions into which the world is divided for frequency allocation purposes. Footnote 533 of the RR urges Administrations to avoid assignments to other stations which may cause harmful interference to radio astronomy within this band. The 25,550–25,600 kHz and the 25,600–25,670 kHz portions of the upper band were allocated to the fixed and mobile (except aeronautical mobile) services and to the broadcasting service, respectively, prior to WARC-79. Fixed and mobile stations were grandfathered into the lower portion of the band (Footnote RR 545), their relocation to be completed by July 1, 1989. This portion of the band should, therefore, be clear of all emissions by now. Relocation of the broadcasting stations assigned to the upper portion of the band was left up to future HF Broadcast Planning Conferences by WARC-79.

Given their increased utilization, radio astronomers are likely to face questions about their continuing need of the HF bands. Knowledge of current and planned uses of these bands, either from the ground or from space, is, therefore, of interest. Intensive use (either active or passive) provides the best argument for the continued allocation of a band to the service using it. Low-frequency radio astronomers should, therefore, register any radio telescopes observing within the HF bands with the International Frequency Registration Board (IFRB) and publicize their plans for future uses. Unless they do so, the HF bands allocted to radio astronomy may face the threat of being reallocated to another service in the future.

References:

[1] Grant, W.B., Thompson, R.E., Haydon, G.W., Steele, F.K., and Cohen, D.J., 1989, NTIA Tech. Memorandum 89-141

III. LOW FREQUENCY SOLAR SYSTEM ASTRONOMY

SOLAR RADIO ASTRONOMY AT LOW FREQUENCIES

George A. Dulk
Department of Astrophysical, Planetary and Atmospheric Sciences
University of Colorado, Boulder

Abstract

Powerful radio radiation often originates from the Sun at decametric and kilometric wavelengths. Radiation from the quiet Sun is produced by the thermal mechanism of bremsstrahlung, and radio bursts of several kinds are produced by the non-thermal mechanisms of plasma radiation and, rarely, gyrosynchrotron radiation.

Radio bursts are frequent during years of high solar activity. They are produced by several mechanisms of excitation: i) streams of electrons, individual, in groups of ten or so, or in storms of tens of thousands, ii) shock waves, with the radiation coming either from near the shock or distant from it by electrons accelerated by the shock, and iii) configurations of magnetic field that contain energetic electrons. In the frequency range of interest here, the radiation is produced at a distance from the Sun that varies from $R \approx 2 \, R_\odot$ at 20 MHz to $R \approx 10 \, R_\odot$ at 1 MHz, and $R \approx 1$ AU at 30 kHz. In that frequency range the observed flux densities range from $S \sim 10^{-17}$ to 10^{-14} W m^{-2} Hz^{-1}, and values of brightness temperature range from $T_b \sim 10^{12}$ to $10^{15.5}$ K.

To the present time there have been almost no observations of solar radiation between about 20 MHz, the limit of observations from the ground, and about 2 MHz where most observations from spacecraft commence. Below about 2 MHz, dynamic spectra of flux densities of solar bursts have been recorded, and, by the use of (short) dipoles on spinning spacecraft, observations have been made of the directions of centroids and characteristic sizes of the emitting sources. Even in the absence of true imaging capabilities, this information has proved to be very valuable in studies of morphology, brightness temperatures, and physics of solar radio phenomena. Proper imaging of the sources will remove present ambiguities and allow new phenomena to be observed.

I. Introduction

In this paper we review solar radio emissions at wavelengths from decametric to kilometric. More general reviews of solar radio emissions include those of Krüger (1979), Kundu (1983), Kundu and Gergely (1980), Dulk (1985, 1990) and Pick (1990). The most extensive, recent review is the book *Solar Radiophysics* edited by McLean and Labrum (1985). Most of these reviews concentrate on emissions at decametric and shorter wavelengths, i.e., emissions that one can observe from Earth.

The principal solar radio observations are of two kinds:

1) With *dynamic spectra* the radiation from the entire Sun is recorded as a function of frequency and time. These observations have been very important for the identification of bursts of different kinds, for example, those that result from electron streams, from shock waves, etc. The reason for the effectiveness is as follows: In general, the lower the frequency of radiation, the farther from the Sun's surface it is emitted. The density n_e and the plasma frequency ν_p decrease with altitude and radiation at frequency ν can only arise from regions where $\nu \gtrsim \nu_p$. Frequently a solar flare originates a perturbation that travels through the corona and the solar wind and produces radiation at or near the plasma frequency and its harmonic. On a dynamic spectrum the radiation is seen to drift from high to low frequencies at a speed that depends on the speed of the perturbation, the rate of decrease of the density of the ambiant electrons, and whether the radiation is at the fundamental or the harmonic of the plasma frequency.

2) With *radioheliographs* (or sometimes simpler instruments) one can observe source locations, dimensions, brightness temperatures and degrees of circular polarization. A few years ago there were radioheliographs at Culgoora in Australia, Clark Lake in California and Nançay in France, where the lowest-frequency images were at 43, 26 and 169 MHz respectively. Today that at Nançay is the only one still operating. At frequencies lower that 26 MHz true images have never been made, but the directivities of dipole or "V" antennas have been used to observe the directions of centroids of sources, characteristic dimensions, and brightness temperatures.

II. The Radiation of the Quiet Sun

The appearance of the quiet Sun changes from frequency to frequency. At high frequencies, $\gtrsim 1$ GHz, it is a well-defined disk superimposed with bright active regions and somewhat dimmer coronal holes. At low frequencies, less than about 100 MHz, the disk is poorly defined, being about $R \gtrsim 1.5$ R$_\odot$, and having a brightness distribution that decreases more or less regularly away from the center. Figure 1 shows some of these characteristics.

At decametric wavelengths, the brighter regions are rarely active regions but are usually coronal condensations lying above filaments. When on the disk, bright regions sometimes lie within or to one side of coronal holes. However, when they are are near the Sun's limb, the radio brightness of coronal holes is depressed (Sheridan and McLean 1985; Kundu et al. 1987). These characteristics imply that the plasma of coronal holes is of low density (giving a depressed brightness when the sources are optically thin, as is true at decimetric wavelengths and above the limb at decametric wavelengths), but perhaps of high temperature (giving enhanced brightness of optically thick sources on the disk at decametric wavelengths). However it might also be argued that the temperature or the density gradient in coronal holes is lower than that of the average quiet Sun, giving an increased optical depth at decametric wavelengths where $\tau \approx 1$ (Sheridan and Dulk 1980).

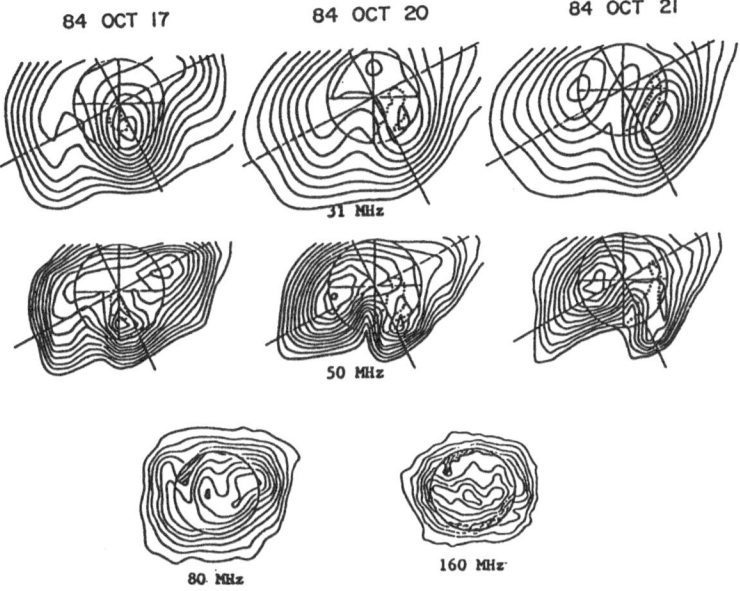

84 OCT 17   84 OCT 20   84 OCT 21

31 MHz

50 MHz

80 MHz   160 MHz

Figure 1. Images of the quiet Sun at several metric and decametric wavelengths as recorded by radioheliographs at Culgoora at 80 and 160 MHz (Sheridan and McLean 1985) and at Clark Lake at 50 and 31 MHz (Kundu et al. 1987). The visible disk of the Sun is shown by the circle. Coronal holes are indicated by dotted lines in the top figures and by cross hatching in the bottom ones.

The flux and brightness temperature of the quiet Sun become more and more difficult to measure at frequencies less than about 100 MHz because the contrast with the bright galactic background becomes less and less. Figure 2 shows measurements of the flux density and of the brightness temperature of disk center at metric and decametric wavelengths. It seems that 26 MHz is the lowest frequency at which these quantities have been measured.

The radiation of the quiet Sun is bremsstrahlung in a layer of plasma above the level where ν_p equals the frequency of observation ν. At frequencies lower than about 100 MHz the brightness temperature of disk center is less than the coronal temperature of $1 - 2 \times 10^6$ K, particularly at frequencies $\lesssim 50$ MHz (Figure 2). The reason for this is not known. Three possibilities have been suggested: i) There is not enough plasma in the layer above the plasma level to make the optical thickness greater than unity (e.g. Smerd 1950). ii) There is a suppression of the emissivity where the index of refraction is less than unity, which is the case at low frequencies where most of the radiation is emitted close to the plasma level (Melrose and Dulk 1988). iii) A large amount of scattering of the radio waves enlarges the source size and decreases its brightness (Aubier, Leblanc and Boischot 1971).

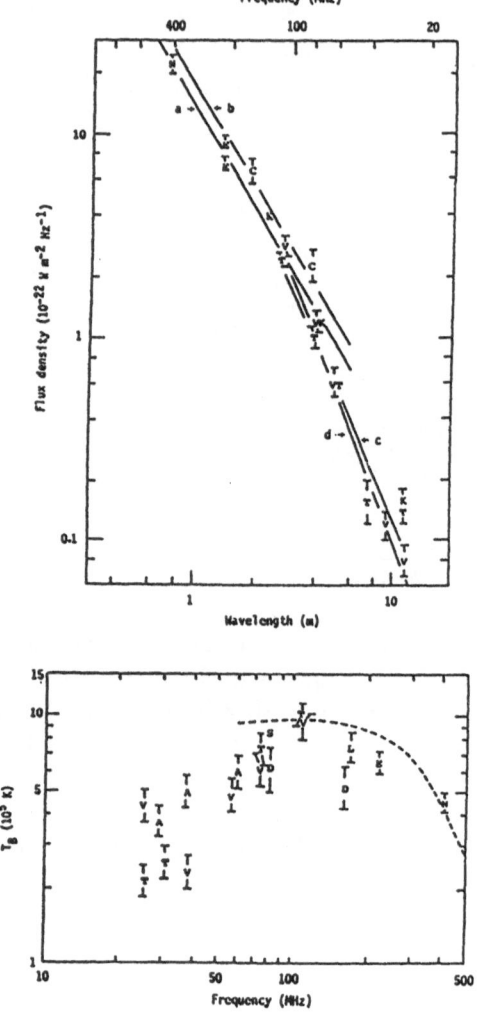

Figure 2. Top: The spectrum of flux density of the quiet Sun from meter to decameter wavelengths. Bottom: The spectrum of brightness temperature of disk center. The dashed line shows theoretical values derived by Smerd (1950) assuming a coronal temperature of 10^6 K. The symbols are explained in the review by Sheridan and McLean (1985), the source of this figure.

III. Non-thermal Radiation from the Active Sun

Radio Bursts of Type III: Electron Beams

Type III radio bursts occur frequently during years of high solar activity. They have been intensely studied, both with observations and theory, because they embody the simplest physics, and a great deal of *in situ* and remote data are available. Type III bursts originate from active regions, most of them in the absence of flares, when packets of electrons are accelerated and then travel along open magnetic field lines through the corona and solar wind at a speed of 0.1 to 0.3 c. As they move along, the faster electrons leave the slower ones behind. At some distance from the region of acceleration, the fast electrons, intermixed with the ambient electrons, form an anisotropic velocity distribution, "bump-on-tail", that is unstable to the production of Langmuir waves. The Langmuir waves grow to a large amplitude, and a part of their energy is converted to electromagnetic waves at the fundamental and second harmonic of the local plasma frequency. All of the components—electrons, plasma waves and radio waves—have been observed by spacecraft in the solar wind (e.g. Lin et al. 1981; 1986; Suzuki and Dulk 1985; Dulk 1990). The spacecraft ISEE–3, which was at the Lagrangian point L1 about 10^6 km sunward of the Earth, made many excellent observations of these various parameters.

Figure 3 shows the relationship between the electrons and the two kinds of waves as observed by ISEE–3. The event is seen to commence with the radio burst at high frequencies that are produced near the Sun. The onset of the radio waves, which is later and later at the lower and lower frequencies, leads directly to the onset of Langmuir waves in the plasma surrounding ISEE–3 that are manifested by the spiky and very intense bursts seen at 30 kHz and weakly at 36 kHz. (The plasma frequency in the vicinity of ISEE–3 was 26 kHz.) It is therefore obvious that the initial radiation of the radio bursts is at the fundamental of the plasma frequency. In addition, we see that the radio radiation at 30 and 36 kHz continues long after the Langmuir waves cease, demonstrating that this late radiation is at the harmonic of the plasma frequency (Dulk et al. 1987).

The top part of Figure 3 shows that most of the electrons of energy $\gtrsim 16$ keV had passed the spacecraft before the onset of the Langmuir waves, and that ≈ 8 keV electrons were arriving. From this we deduce that the electrons that produce radio waves have speeds of about 0.15 c, which is compatible with the time (≈ 80 min) for the 8 keV electrons to traverse the solar wind along the magnetic field, the Archimedean spiral that leads from the Sun to the Earth.

The flux density, the source size and the brightness temperature of type III bursts increases from decametric to kilometric wavelengths. Table 1 shows typical and maximum values of these quantities at several frequencies between 80 and 0.1 MHz. We see that the flux density and brightness temperature attain their maxima near 1 or 2 MHz. Between ≈ 2 and ≈ 40 MHz good measurements of these quantities have never been made.

Type III Bursts from Behind the Sun

Several years ago it was discovered that electron streams produce kilometric type III bursts that are observable by a sensitive receiver (such is on ISEE–3) *irrespective of whether they originate in front of or behind the Sun* (Dulk et al. 1985). This surprising

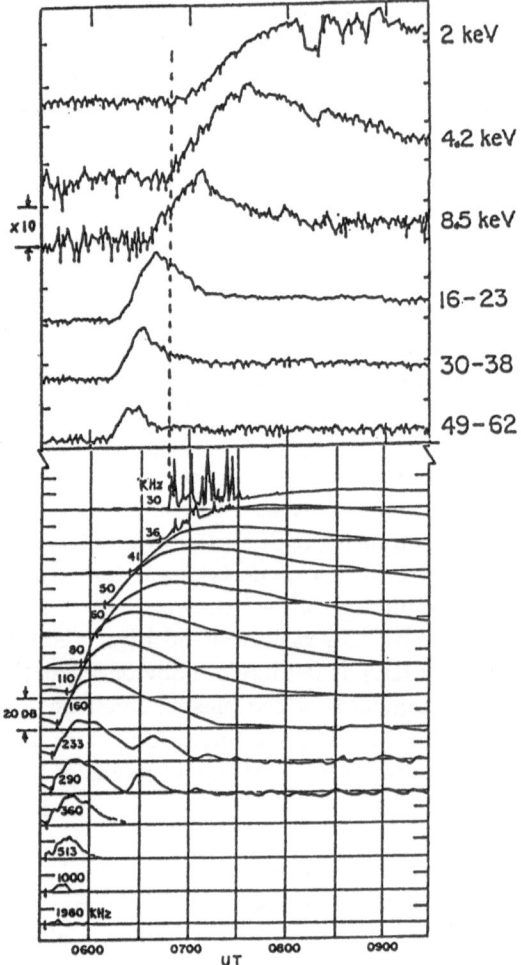

Figure 3. Electron beam, Langmuir wave and type III event of 1979 Feb 8. The plasma frequency at ISEE–3 was $f_{p\oplus} \approx 26$ kHz. The top portion (from Lin et al. 1986) shows the time histories of electrons of several energies arriving at ISEE–3. The bottom portion (from Melrose, Dulk and Cairns 1986) shows the time histories of radio waves recorded by ISEE–3. At 30 kHz (and weakly at 36 kHz) the spiky structures are the product of local Langmuir waves. The dashed line indicates the time of onset of Langmuir waves.

discovery was made by comparing bursts recorded by ISEE–3 at frequencies $\lesssim 2$ MHz with those recorded by Voyager I or II during its travel behind the Sun. The typical flux of bursts arriving from behind the Sun is 20–200 times larger than that of bursts in front of the Sun (Lecacheux et al. 1989). For those few bursts that originate almost directly behind the Sun, the sources appear to surround the Sun like a halo (Reiner and Stone 1990).

Table 1. Typical and maximum values of flux density, brightness temperature and source size of type III bursts.

f (MHz)	$\langle \log S \rangle$ (W m^{-2} Hz^{-1})	$\log S_{max}$ (W m^{-2} Hz^{-1})	$\langle \log T_B \rangle$ (K)	$\log T_{Bmax}$ (K)	Angular Diameter
80	−19.0	−17.0	10.0	11.5	10'
40	−18.5	−16.5	10.5	12.0	20'
2	−15.4	−14.0	14.1	15.5	4°
1	−15.8	−14.0	13.9	15.5	7°
0.3	−16.0	−14.5	13.4	15.0	20°
0.1	−17.0	−15.0	12.5	14.5	70°

The reason for this extended visibility of the radiation is not known in detail, but it is probably due to strong scattering of the radio waves as they traverse the interplanetary medium. There are inhomogeneities of all scales from less than 100 km to several tenths of an AU. Figure 4 shows graphically the great variation in density that is typical: the outer contour of density 44 cm^{-3} ($\nu_p = 60$ kHz) exists at 0.2 AU at some longitudes and at 1 AU at others. The small-scale inhomogeneities are effective in scattering the radio waves, and are particularly effective scatterers when embedded in overdense, large-scale structures because the scattering efficiency is proportional to μ^{-4}, where μ is the index of refraction. The large-scale structures such as sector boundaries refract radio waves (as well as enhancing the scattering), permitting some waves to propagate in directions that would otherwise be forbidden, and preventing other waves from travelling in directions that would otherwise be permitted. Hence, from near the Earth, radiation is recorded that is produced no matter where around the Sun, but the flux may vary from less than 1% to more than 50% of that which one would record in the preferred direction (which, as shown by Reiner and Stone (1989), is outward from the source along the magnetic lines of force).

Storms of Type III Bursts

In addition to the classical type III bursts described above, there are storms of type III bursts that are associated with intense active regions and big sunspots (e.g. Bougeret, Fainberg and Stone 1984; review by Kai, Melrose and Suzuki 1985). A storm may continue for days or weeks and may be composed of more than 10,000 bursts. These decameter/kilometer wavelength type III storms are closely associated with type I storms at metric wavelengths, with the transition between the two occurring at about 50 MHz. Frequently, continuum radiation accompanies the type III storms at decametric wavelengths, but it seems that there is little or no such continuum at frequencies $\lesssim 2$ MHz. At frequencies less than about 0.3 MHz, the duration of individual type III bursts is so long that the bursts overlap and appear like continuum radiation.

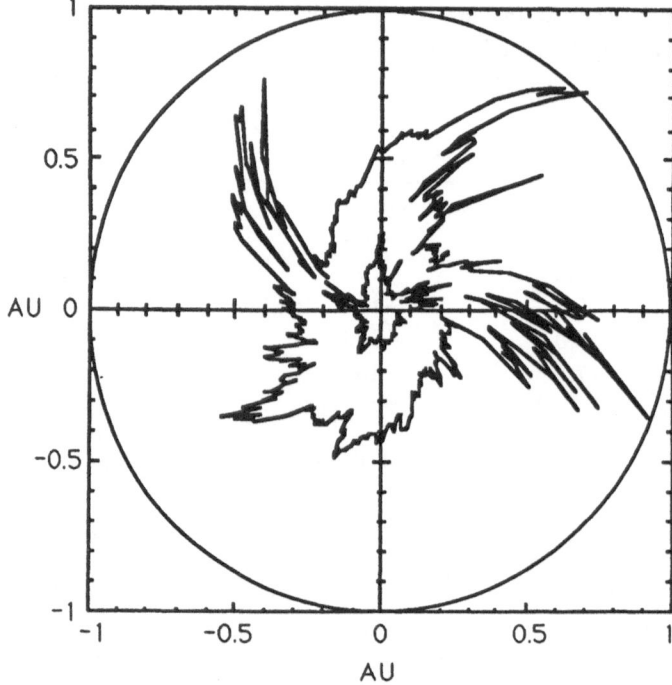

Figure 4. Two contours of constant electron density in the ecliptic plane, the inner one corresponding to f_p = 240 kHz (n_e = 710 cm^{-3}) and the outer one to f_p = 60 kHz (n_e = 44 cm^{-3}). The curves were constructed from 27 days of ion density measurements, one per hour, made between 20 Feb and 18 Mar 1979 (Couzens and King 1986). When constructing the plot, the density was assumed to be proportional to R^{-2} and the Archimedean spirals were constructed using a solar wind speed of 400 km s^{-1}. (From Lecacheux et al. 1989).

Normally the flux and brightness temperature of type III storms are not as high as those of classical type III bursts for which the parameters are given in Table 1.

Flare-Associated Radio Bursts

A number of different kinds of radio bursts are known to be associated with flares, particularly bright and impulsive flares that produce intense radiation throughout the spectrum, but particularly x rays. At meter and decameter wavelengths most of the various types were discovered on dynamic spectra because they produce characteristic patterns of emission as a function of frequency and time (e.g. Wild, Smerd and Weiss

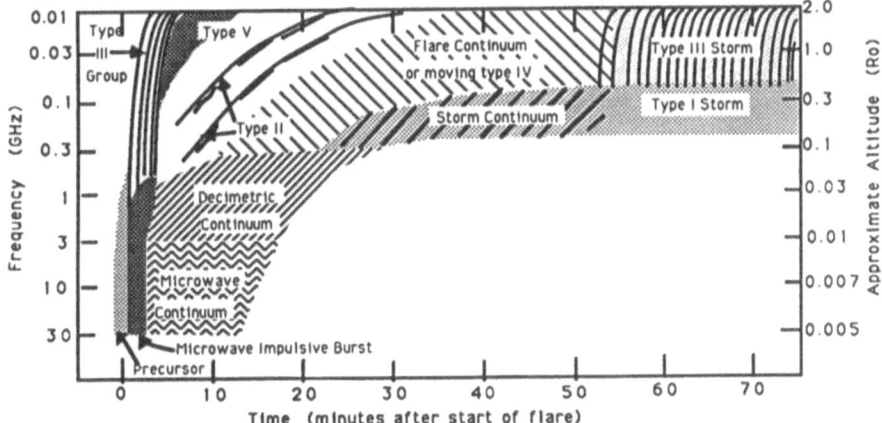

Figure 5. Schematic dynamic spectrum of a solar outburst such as might be produced by a large flare. Individual outbursts can differ from this "typical spectrum", with one or more of the components missing. (Adapted from Dulk 1985).

1963). Later, imaging instruments were used to measure the details of brightness and polarization structure, and to differentiate among various types of continuum bursts.

Figure 5 is a schematic that shows the relationship among the principal kinds of bursts that are associated with major flares. While this represents a typical spectrum, there are immense variations from one flare to another, and not all components occur with all flares. In the following we discuss the kinds of bursts that occur at decametric wavelengths, some of which extend into the realm of 20 MHz to < 1 MHz, of particular interest in the present paper.

Type II Bursts: shock waves

Some solar flares produce shock waves that travel outward through the corona and sometimes the solar wind at speeds of 500 to 1500 km s^{-1}. The shock waves accelerate electrons and produce radio waves, type II bursts, at the fundamental and second harmonic of the plasma frequency (e.g. Nelson and Melrose 1985). The shocks traverse the corona to about $R \approx 2$ R$_\odot$ in about 10–30 min, emitting radio waves that, on dynamic spectra, drift in frequency from ≈ 200 MHz to ≈ 20 MHz (Figure 5). Sometimes these shocks (or others that do not produce radio waves at $\nu \gtrsim 20$ MHz) continue to emit radio waves for 2–3 days while they traverse the solar wind to and beyond 1 AU. During this time the radio waves drift from ≈ 20 MHz to $\lesssim 30$ kHz (e.g. Cane et al. 1982; Kennel et al. 1982). The bursts between ≈ 20 MHz and ≈ 2 MHz have seldom been observed because appropriate recording apparatus has not been available.

The emission mechanism of type II bursts is not well known. It seems that the shock accelerates electrons to energies above a few keV and that these electrons develop an anisotropy that is unstable to the production of Langmuir waves and then radio waves. The type of anisotropy is not known. Sometimes there is evidence of streams of fast electrons that emanate from the shock, but ordinarily there is not.

Shock-Associated Bursts

At frequencies between several MHz and a few hundred kHz, radiation is sometimes observed simultaneously with metric-wavelength type II bursts (Cane and Stone 1984; Kahler, Cliver and Cane 1989). Although the process is poorly known, it may be that the type II shock wave in the corona accelerates electrons that travel some tens of solar radii to the ≈1 MHz plasma level, develop an anisotropy, and emit plasma radiation.

Miscellaneous

There are several types of radiation that one observes at frequencies $\gtrsim 20$ MHz that do not exist or are difficult to identify at frequencies $\lesssim 2$ MHz. Some of them are shown in Figure 5. It seems that they disappear or become very weak somewhere between 20 and 2 MHz.

1) *Flare Continuum* is intense, continuum radiation (Figure 5) that accompanies large flares (e.g. review by Robinson 1985). At 43 MHz brightness temperatures are typically $\sim 10^9$ K, but sometimes reach 10^{12} K. The counterparts of these bursts have never been observed at $\lesssim 2$ MHz.

2) *Storm Continuum* is a type of radiation that occurs in the later phases of large flares at decametric wavelengths (e.g. review by Kai et al. (1985), see Figure 5) and is similar to the type III storms that were described earlier; in fact, the storm continuum following some flares develops into storms that last from days or weeks. A special feature of some storm continuum events is that fine structure bursts are superimposed: *Drift pairs* are double bursts where it appears that the second is an echo of the first, displaced in time by a few seconds, with both components drifting in frequency either toward higher frequency (reverse drift pairs) or toward lower frequency (forward drift pairs). *Fiber bursts* have a very narrow bandwidth ($\ll 1\%$) and drift in frequency at a rate between that of bursts of type II and type III.

The interest in these bursts is largely because of their importance to theory: they have such distinctive and unusual forms that they should help identify particular plasma physical processes that are occurring in the corona and solar wind.

3) *Moving type IV bursts* are associated with sources that originate near the solar surface and move outwards through the corona, sometimes to several $R_\odot$ (e.g. review by Stewart 1985). They are quite rare, only several of them per year. They emit continuum

radiation, but of fairly narrow frequency range, $\lesssim 2{:}1$, with a center frequency that decreases with time. Sometimes they accompany ejections of coronal material, and then the radiation mechanism may be plasma emission because the ejecta may retain an adequately high density for the hour or so required to reach the great heights observed; however the means of containing the needed fast electrons in that ejecta is not clear. In other cases the sources may be plasmoids, closed configurations of plasma and magnetic field that contain fast electrons, where these electrons emit radiation either by the gyrosynchrotron process or by plasma radiation.

It is not known to what lower frequency limit these various kinds of bursts continue because there have been no observations between ≈ 20 and ≈ 2 MHz.

IV. Conclusions

The frequency band between ≈ 20 and ≈ 2 MHz is effectively unexplored. No true images of solar sources have ever been made at frequencies less than 26 MHz. In the 1990's the spacecraft WIND will make spectral observations in the range from ≈ 16 MHz to < 10 kHz. With spinning dipole antennas, it will not make true images but will be able to measure the directions of centroids of sources, their characteristic dimensions, and hence their brightness temperatures.

In the discussion above we have not described the polarization of the various bursts. At frequencies $\gtrsim 20$ MHz the degree of circular polarization of type III bursts is often 10%–40% and that of some kinds of continuum radiation is $\approx 100\%$. (There is no linear polarization in solar bursts because of depolarization by the Faraday effect.) At frequencies lower than 2 MHz a significant degree of polarization has never been observed. Hence it is important to determine how the polarization evolves in the band from 20 to 2 MHz. This would furnish independent information that is almost equally important as that of spectra of flux and images of sources.

Acknowledgements: I thank Dr. J.L. Steinberg for his valuable comments This work has been supported in part by NASA, Grants NSG-7287 and NAGW-91, and by NSF, Grant ATM-8719371, to the University of Colorado.

References

Aubier, M., Leblanc, Y. and Boischot, A., 1971, *Astron. Astrophys.*, **12**, 435.

Bougeret, J.L., Fainberg, J. and Stone, R.G., 1984, *Astron. Astrophys.*, **136**, 255.

Cane, H.V. and Stone, R.G., 1984, *Astrophys. J.*, **282**, 339.

Cane, H.V., Stone, R.G., Fainberg, J., Steinberg, J.L. and Hoang, S., 1982, *Solar Phys.*, **78**, 187.

Couzens, D.A. and King, J.H., 1986, *Interplanetary Medium Data Book, Suppl. 3 and 3a*, NSSDC-WDC-A 86-04 and 86-04a, NASA Goddard Space Flight Center, Greenbelt, MD, USA.

Dulk, G.A., Steinberg, J.L., Lecacheux, A., Hoang, S. and McDowall R.J., 1985, *Astron. Astrophys. (Letters)*, **150**, 28–30.

Dulk, G.A., 1985, *Ann. Rev. Astron. Astrophys.*, **23**, 169.

Dulk, G.A., Steinberg, J.L., Hoang, S. and Goldman, M.V., 1987, *Astron. Astrophys.*, **173**, 366.

Dulk, G.A., 1990, *Solar Phys.*, in press.

Kahler, S.W., Cliver, E.W. and Cane, H.V., 1989, *Solar Phys.*, **120**, 393.

Kai, K., Melrose, D.B. and Suzuki, S., 1985, in *Solar Radiophysics*, eds. D.J. McLean and N.R. Labrum (Cambridge: Cambridge University Press), p. 385.

Kennel, C.F., Scarf, F.L., Coroniti, F.V., Smith, E.J. and Gurnett, D.A., 1982, *J. Geophys. Res.*, **87**, 17.

Krüger, A., 1979, *Introduction to Solar Radio Astronomy and Radio Physics*, (Dordrecht: Reidel), 330 pp.

Kundu, M.R., 1983, *Solar Phys.*, **86**, 205.

Kundu, M.R. and Gergely, E. eds., 1980, *Radio Physics of the Sun, IAU Symp. No. 86*, (Dordrecht: Reidel), 475 pp.

Kundu, M.R., Gergely, E., Schmahl, E.J., Szabo, A., Loiacono, R. and Wang, Z., 1987, *Solar Phys.*, **108**, 113.

Lecacheux, A., Steinberg, J.L., Hoang, S. and Dulk, G.A., 1989, *Astron. Astrophys.*, **217**, 237.

Lin, R.P., Potter, D.W., Gurnett, D.A. and Scarf, F.L., 1981, *Astrophys. J.*, **251**, 364.

Lin, R.P., Levedahl, W.K., Lotko, W., Gurnett, D.A. and Scarf, F.L., 1986, *Astrophys. J.*, **308**, 954.

McLean, D.J. and Labrum, N.R. eds., 1985, *Solar Radiophysics*, (Cambridge: Cambridge University Press), 516 pp.

Melrose, D.B., Dulk, G.A. and Cairns, I.H., 1986, *Astron. Astrophys.*, **163**, 229.

Melrose, D.B. and Dulk, G.A., 1988, *Solar Phys.*, **116**, 141.

Nelson, G.J. and Melrose, D.B., 1985, in *Solar Radiophysics*, eds. D.J. McLean and N.R. Labrum (Cambridge: Cambridge University Press), p. 333.

Pick, M., 1990, *Solar Phys.*, in press.

Reiner, M.J. and Stone, R.G., 1989, *Astron. Astrophys.*, **217**, 251.

Reiner, M.J. and Stone, R.G., 1990, *Astron. Astrophys.*, in press.

Robinson, R.D., 1985, in *Solar Radiophysics*, eds. D.J. McLean and N.R. Labrum (Cambridge: Cambridge University Press), p. 385.

Sheridan, K.V. and Dulk, G.A., 1980, in *Solar and Interplanetary Dynamics*, eds. M. Dryer and E. Tandberg-Hanssen, (Dordrect: Reidel), p. 37.

Sheridan, K.V. and McLean, D.J., 1985, in *Solar Radiophysics*, eds. D.J. McLean and N.R. Labrum (Cambridge: Cambridge University Press), p. 443.

Smerd, S.F., 1950, *Austr. J. Sci. Res., Ser. A*, **3**, 34.

Stewart, R.T., 1985, in *Solar Radiophysics*, eds. D.J. McLean and N.R. Labrum (Cambridge: Cambridge University Press), p. 361.

Suzuki, S. and Dulk, G.A., 1985, in *Solar Radiophysics*, eds. D.J. McLean and N.R. Labrum (Cambridge: Cambridge University Press), p. 289.

Wild, J.P., Smerd, S.F. and Weiss, A.A., 1963, *Ann. Rev. Astron. Astrophys.*, **1**, 291.

SOME PROBLEMS IN LOW FREQUENCY SOLAR RADIO PHYSICS

N. Gopalswamy and M.R. Kundu
Astronomy Program, University of Maryland
College Park, MD 20742

Abstract

Several important problems in solar radio physics can be attacked using the high spatial resolution observations from a low frequency space array, as the problem of ionospheric refraction does not exist. Noise storms are believed to occur in closed magnetic loops due to trapped superthermal particles. Recent radioheliograph observations suggest such a magnetic field topology up to altitudes of about 40 MHz emission. The problem of relative locations and sources of the storm continuum and bursts can be effectively studied by imaging them with higher spatial resolution. Interplanetary type II bursts are observed from heights above ~ 10 $R_\odot$ while coronal type II bursts are observed from heights less than ~ 3 $R_\odot$. Observations filling this gap have important implications for the understanding of solar-terrestrial relations through shocks and mass ejections.

1. Introduction

The Sun can not be observed at frequencies below ~ 20 MHz from ground because of the ionosphere. This corresponds to a radial height of ~ 3-4 $R_\odot$ for bursts due to a plasma emission mechanism where the plasma frequency is determined by the background coronal density. Moving type IV bursts can be observed to greater heights from dense structures or closed magnetic arches containing nonthermal electrons ([1-2]). The two radioheliographs (Culgoora [3] and Clark Lake [4]) imaged the Sun down to 43 MHz and 25 MHz respectively, but they are not operating any more. Although these instruments had relatively poor spatial resolution, they advanced our understanding of the outer structure of the corona considerably. Study of the radio phenomena at lower frequencies has to be done by instruments placed above the earth's ionosphere. The ISEE-3 Solar Radio Experiment observed the Sun in the frequency range 2 MHz - 30 kHz corresponding to emission in the height range 10-215 $R_\odot$ [5]. Thus, there is virtually no information in the height range ~ 4-10 $R_\odot$. This unexplored height range may be the region where the solar wind originates and the magnetic field configuration of the corona changes from closed predominantly to open. In this paper, we discuss some of the issues which emerge from the meter-decameter studies of the Sun which could probably be settled by acquiring imaging data at frequencies below ~ 25 MHz.

2. Storm Radiation

Solar noise storms start at ~ 300-500 MHz and are observed down to ~ 40 MHz below which it is believed that they change character to become what is known as decmetric type III storms. Dynamic spectra show an abrupt transition from type I to type III storms. The type III storms seem to continue into the interplanetary medium. Fainberg and Stone [6] have also detected a continuum component in the hectometer wavelength range. Storms at all these height ranges have a smooth continuum with superposed type I bursts in the metric frequencies and type III bursts in the decameter and hectometer domains. The metric continuum is believed to be due to superthermal electrons trapped in closed magnetic arches which produce a steady level of Langmuir waves. The Langmuir waves combine with some low frequency waves such as ion-sound or lower hybrid waves to produce the observable radiation [7]. Local enhancements in Langmuir wave densities result in the type I bursts. Decametric and hectometric storms are believed to occur in open field lines, with the electrons that escape into the open field lines producing the type III bursts. It is easy to understand the trapping and storage of superthermal electrons in closed arches; however, it is somewhat difficult to understand the production of continuum when the electrons propagate along open field lines [8]. Knowledge of the low frequency spectrum and source structure has important implications for the theory of storm radiation, e.g., in accounting for the transition from closed to open field lines.

Fig. 1: The sizes of continuum and burst components of August 3, 1985 storm. The beam size is indicated.

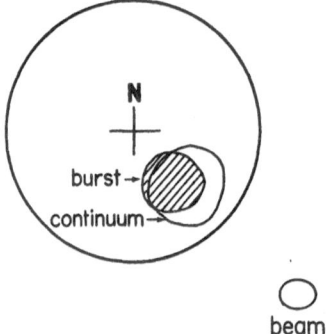

From radioheliograph observations, it has been found that the type I bursts and continuum at a given frequency have the same source location (see Fig. 1) with the bursts having a smaller size [9]. Sometimes more than one burst source is found to be located in a continuum source [10]. Observations indicate that the source sizes of type I bursts are ~ 2-3' at 100-200 MHz increasing at lower frequencies. Lowest size observed is 0.'7 [11]. It is important to determine the size of the bursts for understanding the emission mechanism. Short rise times (~ 0.1s) suggest a size $\leq 10^3$ km; absence of second harmonic emission suggests $\gtrsim 10^3$ km [7]. Accurate estimate of the size will lead to a better understanding of the emission mechanism. Similar argument applies to fine structures such as drift pairs in the type III storms. Although the type I-type III transition take place around 40 MHz, it may be possible to observe a small fraction of type I bursts at lower frequencies and determine the extent of closed field lines in the corona.

3. Origin of Type II Shocks

A clear causal relation was expected between cornal mass ejections (CMEs) and type II bursts during the Skylab era because only high speed CMEs were associated with type II bursts. It was believed that the CME moving at super Alfvenic speed piston drove a shock responsible for the type II burst. The shock should then be at the "stand-off" distance ahead of the CME leading edge and hence the type II location should be radially above the CME in the plane of the sky. However, when simultaneous CME and radio imaging observations were available during the SMM era, the type II burst location was rarely found ahead of the CME leading edge [12-13]. In one event (June 27, 1984), although the type II location was ahead of the CME, the CME was so small (20 times smaller than the inferred shock extent) that the CME could not have piston driven the shock [14]; the shock was interpreted to be due to a flare explosion (blast wave). Fig. 2 shows two examples of simultaneous white light and

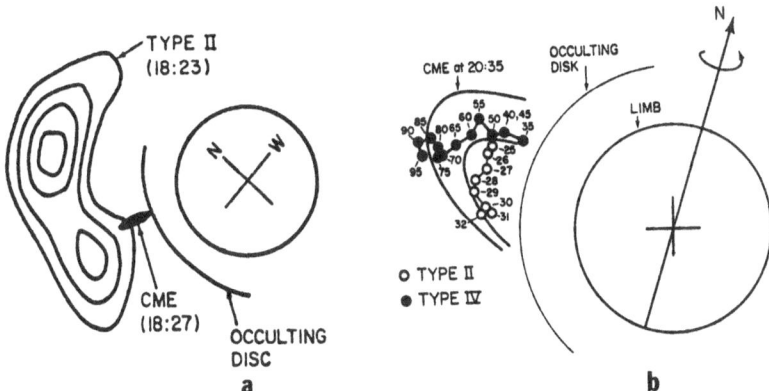

Fig. 2: a) Clark Lake type II burst contours superposed on SMM-C/P image of a CME [14]. b) Superposition of type II and moving type IV centroids on SMM-C/P CME [13].

radio observations where the relative positions of CMEs and type II bursts are indicated. Therefore, both blast wave and piston driven shocks can in principle generate a type II burst, although there is no observational evidence for the latter. The observation of a type II burst may depend on several factors: (i) the height of shock formation from the Sun-center; (ii) the distance the shock propagates before being dissipated. The item (i) can be found from the starting frequency in the dynamic spectra of type II bursts. The item (ii) can be found also from the frequency extent of type II bursts in the dynamic spectra. Shocks propagating over large distances can produce both coronal and interplanetary (IP) type II bursts. Robinson and his co-workers [15] found that the type II bursts associated with IP events had starting frequencies less than ~ 45 MHz. IP shocks are also known to be associated with non-impulsive events such as disappearing filaments and eruptive prominences [16-17]. It is possible that the piston driven shocks form at higher altitudes in the corona (corresponding to frequencies below 45 MHz) and continue into the IP medium. On the other hand a blast wave with a slower energy input at its origin, may not develop into a shock in the corona, but may eventually steepen and become a shock. Therefore, what one needs is the relative location between CMEs and spatially resolved type II bursts below 45 MHz. The lowest frequency at which Culgoora radioheliograph imaged type II bursts was only 43 MHz; the Clark Lake radioheliograph has observed type II bursts at 30 MHz [14]. Therefore, very little imaging is available in the range below ~ 45MHz where one expects piston driven type II shocks: The question of blast wave or piston driven shocks for type II bursts may be resolved only when type II bursts can be imaged below this frequency range. Ideally simultaneous white light observations will also be needed.

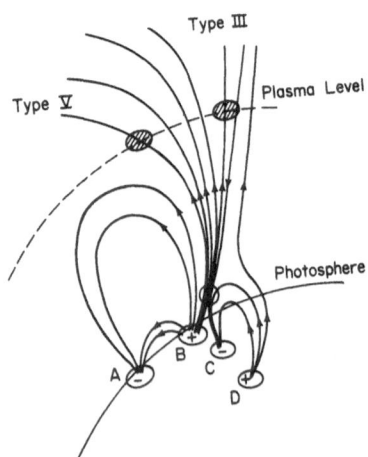

Fig. 3: Relative location of type III and type V bursts during February 3, 1986 flare [18].

4. Type III-type V Bursts

There is a separate class of type III bursts ("normal type III bursts"), of different origin than the storm type III bursts. These are often associated with flares, but can also occur due to minor impulsive events. During flares, the type III bursts can be followed by relatively longer duration emission called type V bursts. These type III and type V bursts are believed to be due to relativistic electron beams propagating along open or diverging magnetic fields (see Fig. 3). Type III-type V bursts have been observed at decametric wavelengths at Clark Lake [18] and it will be interesting to find up to what height one sees neighboring closed and open field lines above flaring active regions.

In conclusion, low frequency radio observations of solar radio phenomena in the frequency range 1-30 MHz will confirm existing ideas based on meter-decameter wavelength observations. In addition they will improve our understanding of the transition of many physical phenomena from the corona to the interplanetary space.

Acknowledgements

This research was supported by NSF grant ATM 88-16008 and by funds obtained from the University of Maryland Astronomy Program, the College of Computer, Mathematical and Physical Sciences, and the Office of the Dean of Graduate Studies and Research.

References

1. Sheridan, K.V.: 1970, Proc. Astron. Soc. Australia 1, 138.
2. Gopalswamy, N. and Kundu, M.R.: 1989, Solar Phys. 122, 145.
3. Wild, J.P.: 1967, Proc. IREE Australia 9, 279.
4. Kundu, M.R., Erickson, W.C., Gergely, T.E., Mahoney, M.J., and Turner, P.J.: 1983, Solar Phys. 83, 365.
5. Stone, R.G., Cane, H.V., and Bougeret, J.-L.: 1984 in STIP Symposium on Solar Interplanetary Intervals, eds. M.A. Shea et al., pp.371-382, Engineering International, Huntsville.
6. Fainberg, J. and Stone, R.G.: 1971, Astrophys. J. 164, L123.
7. Melrose, D.B.: 1980, Solar Phys. 67, 357.
8. Levin, B.N.: 1982, Astron. Astrophys. 111, 71.
9. Suzuki, S.: 1961, Ann. Tokyo Astron. Obs. 7, 75; Kundu, M.R. and Gopalswamy, N.: 1989, Solar Phys. (in press).
10. Daigne, G.: 1968, Nature 220, 567.
11. Kerdraon, A.: 1979, Astron. Astrophys. 71, 266.
12. Wagner, W.J.: 1984, Ann. Rev. Astron. Astrophys. 22, 267.
13. Kundu, M.R., Gopalswamy, N., White, S., Cargill, P., Schmahl, E.J. and Hildner, E.: 1989, Astrophys. J. 347, 505.
14. Gopalswamy, N. and Kundu, M.R.: 1987, Solar Phys. 114, 347.
15. Robinson, R.D., Stewart, R.T., and Cane, H.V.: 1984, Solar Phys. 91, 159.
16. Jocolyn, J.A. and McIntosh, P.I.: 1981, J. Geophys. Res. 86, 4555.
17. Wright, C.S. and McNamara, L.F.: 1983, Solar Phys. 87, 401.
18. Gopalswamy, N. and Kundu, M.R.: 1987, Solar Phys. 111, 347.

BROAD-BAND IMAGES OF AKR FROM ISEE-3

Bernard V. Jackson
Center for Astrophysics and Space Sciences, UCSD
La Jolla, California 92093
and
Jean-Louis Steinberg
Department de Recherche Spatiale, UA CNRS No 274,
Observatoire de Paris, 92195 Meudon Principal Cedex, France

Recent ISEE observations (Steinberg et al., 1989) have shown that radio waves from naturally-occurring auroral kilometric radiation (AKR) in the range of 45 to 150 kHz can in some instances be refracted or ducted away from the location of the origin of the radio source. It is shown that in certain cases the *apparent* position of radiation when viewed at a distance emanates from a position very distant from the Earth. Figure 1

Figure 1:

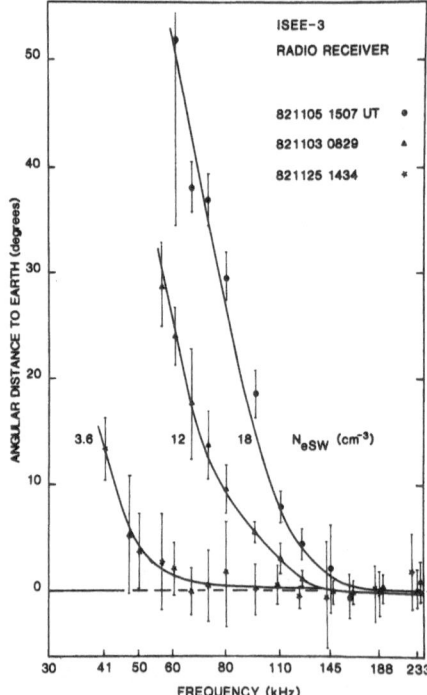

Angular distance to the Earth of three AKR sources observed when ISEE-3 was in the solar wind. The angle increases when the frequency decreases and when the solar wind density N_{eSW} increases.

(from Steinberg et al., 1988) shows the location of the apparent source of AKR observed from ISEE-3 at different frequencies on three different days when the solar wind density had different values. At the time ISEE-3 was well outside the Earth's magnetosheath. Figure 1 shows that, as seen from ISEE-3, the apparent location of the radiation from the spacecraft can be many degrees tailward of the Earth. This implies that the radiation is refracted or ducted along the magnetotail/magnetosheath region until it escapes in the direction of the ISEE-3 spacecraft.

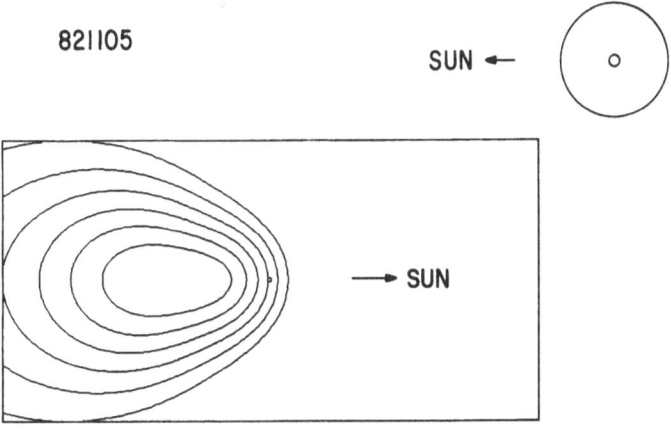

Figure 2:

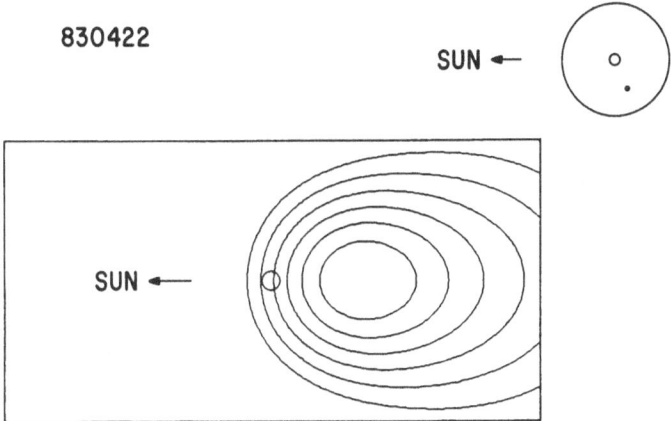

Linear contour images of AKR source intensity at different frequencies from ISEE-3 on the dates indicated. The rectangular image is the view from the spacecraft where equal angular positions are represented by equal linear distances. The rectangle is 50° by 100° in size. The circle at the center of the image depicts the size and location of the Earth. The circle at the upper right hand corner of each image depicts the relative locations of the Earth, Lunar orbit and spacecraft relative to the Sun at the time indicated.

It is clear from the work of Steinberg that this naturally-occurring process can be used to produce images of AKR which propagates along and through the magnetosheath/magnetotail and be used to infer the overall shape of region. Figure 2 is a simple contour plot "linear" broad-band kilometric image of the data in figure 1 for day 821105 and for day 830422 (see appendix I). While it is not known at this time how well the radiation would highlight various features in the region that displace the radiation from its original source, the technique is clearly one that will allow information to be derived about the overall structure of the magnetosheath/magnetotail region of Earth when AKR is active. For instance, according to some theories (ie, Hones, 1979) during a substorm, it is possible that the Earth's magnetotail decreases in length by nearly an order of magnitude. I would expect a less distant displacement of AKR from the Earth when this occurs. As support of *in-situ* observations, the kilometric-wave technique would provide an extremely sensitive mechanism for obtaining an overall global depiction of the magnetosphere, and might show how the magnetotail behaves during a substorm.

References

Steinberg, J.L., S. Hoang and C. Lacombe, Propagation of Terrestrial Kilometric Radiation through the Magnetosheath: ISEE-3 Observations, *Annales Geophysicae*, **7**, 151, 1989.

Hones, E.W., Jr., Transient Phenomena in the Magnetotail and Their Relation to Substorms, *Space Science Revs.*, **23**, 393, 1979.

APPENDIX I

How the Broad-Band Linear AKR Images Were Produced

The linear magnetospheric images were produced using the spin-modulated signal of the ISEE-3 spacecraft as it looped through the Earth-Moon system prior to departing the Earth vicinity for comet Giacobini-Zinner. The kilometer-wave source intensity at a given frequency is spin modulated by the dipole antenna receptor located in the spin plane of the spacecraft. A source intensity null occurs when the antenna is directed toward it. The depth of the intensity modulation is related to the size of the source; the larger sources modulate the intensity least. The spin plane of the ISEE-3 spacecraft lay approximately parallel to the ecliptic plane. Lower frequency sources appear farther from the Earth as is indicated in the literature.

For these linear contour images, I model the intensity from the source at each frequency as if it had a circular gaussian intensity pattern whose size was given by the spin modulation. Furthermore, an ad-hoc model of the source maximum intensity shows that it decreases exponetially with angular distance from the Earth. The exponent was found to be -deg/5.5. With this exponential factor folded into the intensity scale and a further removal of all signal source centers ducted less than 3° from Earth, the intensity signal centered at each frequency observed was accumulated on a rectangular grid of dimension 51×101 and summed. These intensities were then contoured in arbitrary units linearly upward from the 40% level (which represents the lowest level) in steps of 10% to the maximum (100%). The Earth was centered in the grid. This technique essentially forms a broad-band image of the magnetosphere at kilometric wavelengths. Note that this technique gives no directional information and little size information about the source intensity perpendicular to the ecliptic plane. It for this reason that these images are described as *linear* images.

LOW FREQUENCY PROPAGATION
IN THE EARTH'S MAGNETOSPHERE

Brian Dennison
Physics Department, Virginia Tech
Blacksburg, Virginia 24061, USA

S. Ananthakrishnan
Radio Astronomy Center
Tata Institute for Fundamental Research, Bombay, India

M. Desch and M. L. Kaiser
Laboratory for Extraterrestrial Physics
NASA — Goddard Space Flight Center, Greenbelt, MD 20771, USA

K. W. Weiler
Center for Advanced Space Sensing
Naval Research Laboratory, Washington, DC 20375, USA

I. INTRODUCTION

The low frequency space array (LFSA) mission concept (Dennison et al. 1986; Weiler et al. 1988, 1990) is driven by the need to observe above the ionosphere, particularly at frequencies below about 10 MHz, to which the Earth's ionosphere is frequently opaque. In near-Earth orbit, however, such an array will operate within the magnetosphere. In this paper, we report on preliminary estimates of the effect of this medium upon the imaging potential of the LFSA.

II. PROPAGATION MODEL

The relevant geometry is shown in Figure 1. For simplicity, we consider cases in which the wavevector, $\vec{k}$, and the baseline vector, $\vec{B}$ both lie in a common (magnetic) meridional plane at the array. In addition, $\vec{B}$ is assumed to be perpendicular to $\vec{r}$, the radius vector from the center of the Earth. The fringe phase is given by

$$\Phi = \vec{k} \cdot \vec{B} = n \frac{\omega B}{c} \sin \beta, \tag{1}$$

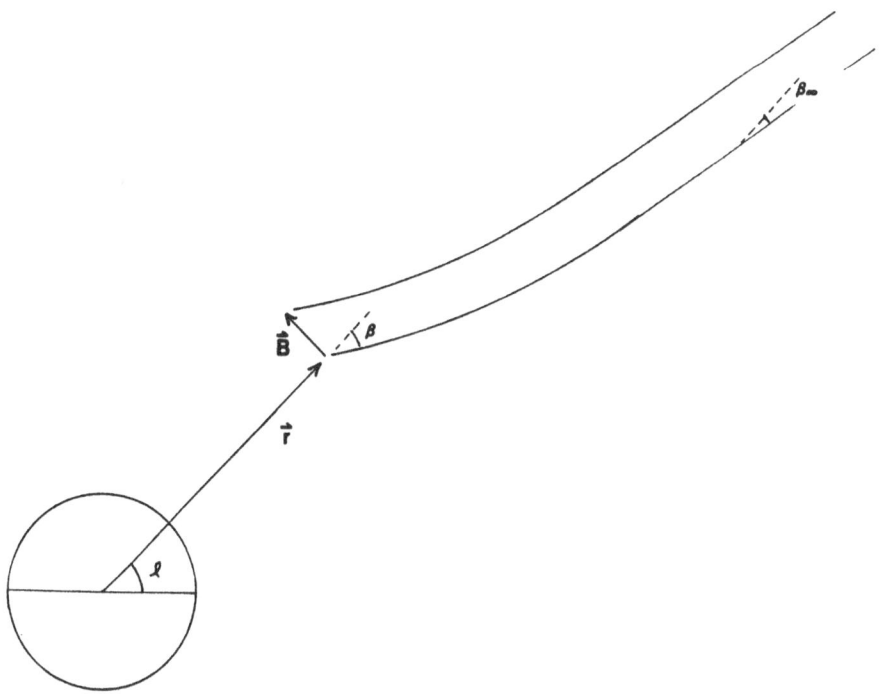

Figure 1. Geometry for wave propagation confined to a meridional plane. The horizontal line represents the Earth's magnetic equator. At low frequencies the useable array dimensions are quite small, $\approx$ 30 km; in which case the ray trajectories are nearly parallel, and the propagation effects can be determined from a single ray trajectory launched from the array. The baseline vector ($\vec{B}$) is exaggerated in this illustration.

where n is the refractive index at the array. The zenith angle of a source at the array, β, differs from the true zenith angle, β_∞ because of global refraction in the magnetosphere. (Both β and β_∞ are referred to the zenith direction established at the array. The source distance is assumed to be very large in comparison with the size of the magnetosphere.) In the absence of a magnetosphere, the fringe phase would be

$$\Phi_0 = \frac{\omega B}{c} \sin \beta_\infty. \tag{2}$$

The shift in fringe phase attributable to the magnetosphere can be expressed as the sum of a local and a global term,

$$\Delta\Phi = \Phi - \Phi_0 = (n-1)\frac{\omega B}{c} \sin \beta + \frac{\omega B}{c}(\sin \beta - \sin \beta_\infty). \tag{3}$$

The first term represents the phase shift due to the presence of ambient medium along the geometric delay, a local effect; while the second term represents the effects of refraction occuring over the full path through the magnetosphere, a global effect. This approach assumes that the rays reaching the two antennas have suffered essentially the same global refraction, that the refractive index is nearly constant over the geometric delay, and that negligible refraction occurs along the geometric delay. These conditions will probably be met for the relatively short baselines under consideration for the LFSA.

In a plane-parallel atmosphere, the global refraction can be calculated simply by applying Snell's Law to an arbitrarily large number of layers, the lowest with refractive index, n, and the highest with unit refractive index, with the result that $\sin \beta_\infty = n \sin \beta$. This yields the familiar result for a plane parallel atmosphere, namely that $\Delta \Phi = 0$. Although the magnetosphere is not plane-parallel, we may in some cases expect a degree of partial cancellation of the terms in equation (3).

The net phase shift was computed for a variety situations thought to be relevant. The local term was easily computed using the general form (Lang, 1980) for the refractive index in a magnetized plasma,

$$n^2 = 1 - \frac{X}{\frac{1-\frac{1}{2}Y^2 \sin^2 \Omega}{1-X} \pm \left[\frac{\frac{1}{4}Y^4 \sin^4 \Omega}{(1-X)^2} + Y^2 \cos^2 \Omega \right]^{1/2}}, \tag{4}$$

where $X = (\omega_P/\omega)^2$, and $Y = \omega_H/\omega$. The angle between the magnetic field direction and $\vec{k}$ is Ω, and ω_P and ω_H are the plasma- and gyro-frequencies. The medium is birefringent, with the plus sign corresponding to the ordinary (O) mode of propagation, and the minus sign to extraordinary (E) mode.

The global term was computed using a Department of Commerce ray tracing program, suitably modified for planetary magnetosphere calculations. In all cases, the ray was launched at the array at various zenith angles (β), and β_∞ was determined at large distance from the Earth.

For these preliminary calculations the magnetosphere was modelled with a simple dipole field and a radial density distribution (Chiu et al. 1979), using equitorial plane density measurements appropriate for solar maximum (Kasha 1967, NASA 1971). With this simple model, calculations were performed for a variety of radii, magnetic latitudes, and zenith angles, and at the lowest altitudes for both fundamental modes (O and E). For all calculations, a frequency of 1.5 MHz was used. Because the medium is birefringent, we find different values for $\Delta \Phi$ in the two modes. The difference permits an estimate to be made of the effects of ray splitting due to birefringence.

Figure 2 shows the results of our ray tracing for the particular case in which the array is in the magnetic equator with $r = 1.5$ R$_\oplus$. Clearly, at this radius the refraction and birefringence are severe. Not suprisingly, we find these effects to be smaller at higher altitudes. The major factors contributing to refraction are the density gradient transverse to the line of sight, and the magnetic field geometry. That is, the refraction is large at large zenith angles, and when the propagation is approximately longitudinal. The latter phenomenon can lead to significant ray bending near the zenith at high magnetic latitudes.

For interferometric observations the quantity indicative of the degree of image distortion is, of course, the net phase shift, $\Delta \Phi$. In Figures 3 and 4, we plot the absolute magnitudes of the global and local phase shifts for the various cases considered. In all

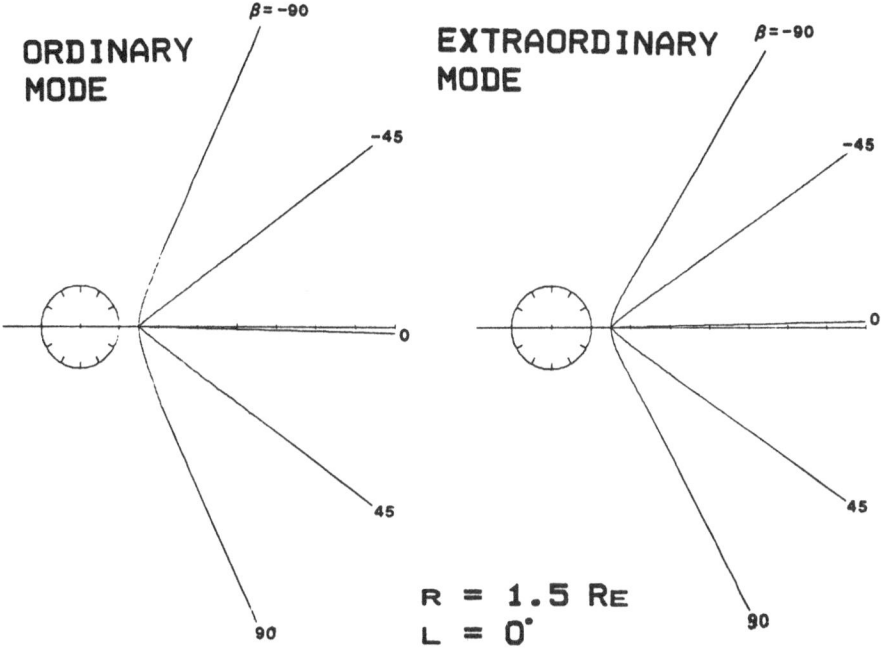

Figure 2. Ray trajectories in the magnetosphere at 1.5 MHz. Each ray is labelled by its zenith angle *at the array* (β).

cases the global and local phase shifts are of opposite sign, and therefore the vertical lines connecting them represent the magnitude of $\Delta\Phi$. Comparison of O and E modes (both computed for only $r = 1.5$ R$_\oplus$) gives the magnitude of the phase shift due to birefringence.

III. RESULTS AND DISCUSSION

From Figures 3 and 4 we arrive at the following results:

1. The phase distortion is very serious for $r = 1.5$ R$_\oplus$. Although the self-correcting tendency of the local and global terms to partially cancel is present, we are still left with a net phase shift of 1 rad km^{-1}, which is prohibitively large.
2. Propagation effects tend to be considerably worse at higher magnetic latitudes, due to large refraction as longitudinal propagation is approached.

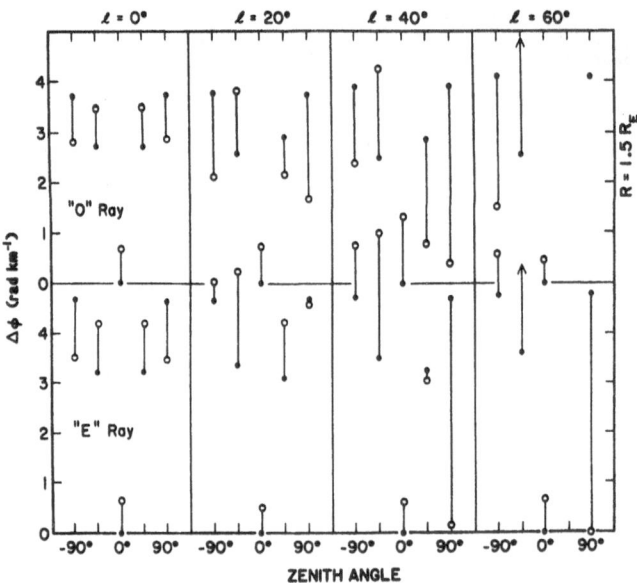

Figure 3. Absolute vaules of fringe phase shifts for $r = 1.5$ R$_\oplus$. Dots represent the local term, and circles represent the global term. Since in all cases the local and global terms are of opposite sign, the lengths of the lines connecting each pair are indicative of the magnitude of $\Delta\Phi$.

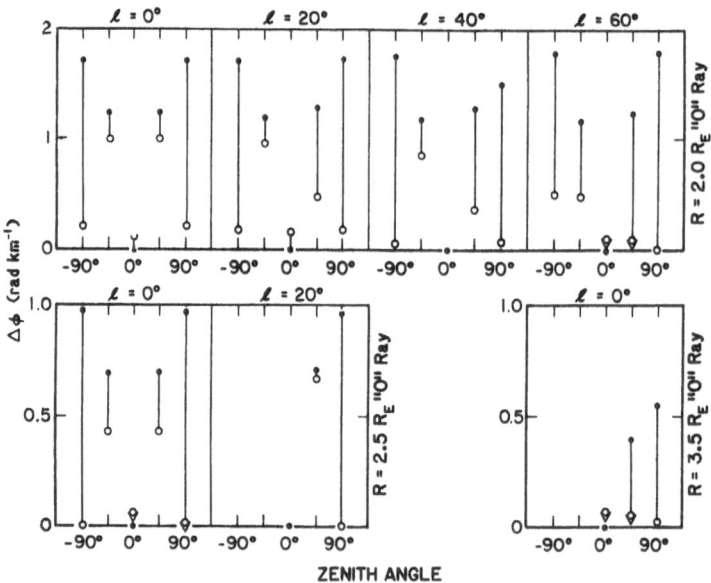

Figure 4. Absolute values of fringe phase shifts for $r = 2.0$, 2.5, and 3.5 R$_\oplus$. Symbols have the same meaning as in Figure 3.

3. Not suprisingly, the phase shifts in the O and E modes tend to follow one another, with the E mode phase shifts being typically 20 percent larger at $r = 1.5\,R_\oplus$. Hence, the birefringent splitting at this radius is about a fifth of the overall displacement.

4. At larger radii, the effects are smaller, as expected. In these cases, however, the departures from plane-parallelism are even greater, and hence there tends to be less cancellation in summing the local and global terms. At $r = 2.5\,R_\oplus$, there remains an appreciable net phase shift at zenith angles of $45°$, i.e. about 0.25 rad km^{-1}. Hence one full turn of net phase shift would occur on baselines ≈ 25 km. Interplanetary and interstellar scattering are expected to yield fringe phase fluctuation of typically a few radians on this baseline. Hence, at this radius magnetospheric distortions are somewhat more severe than the degradation due to scattering which ultimately limits the resolution. At $r = 3.5\,R_\oplus$, the global phase shifts are quite small; however about 0.4 rad km^{-1} of local phase remains uncancelled at $\beta = 45°$.

Overcoming the phase shifts remaining at large radii will hinge upon the feasibility of correcting the data. (See below.) Because of the uncertainties involved, as well as the preliminary nature of our calculations, setting a minimum radius for successful imaging at 1.5 MHz would be premature. The available results suggest, however, that it would be beyond 2.5 $R_\oplus$. Clearly, further study is required. It may be that observations at these frequencies will need to be carried out exterior to the plasmapause (at $r \approx 4\,R_\oplus$). The birefringent splitting also needs to be studied at radii larger than 1.5 $R_\oplus$, although we expect that it will be a smaller fraction of the overall effect than the 20 percent found at $r = 1.5\,R_\oplus$.

It should also be pointed out that the calculations described above were carried out in the monochromatic limit. For finite bandwidths smearing of images will also occur.

Correcting the data for the above-mentioned effects might be possible, but it won't be easy. This is because the correction depends upon time, direction, frequency and polarization. It is important to bear in mind that the needed correction will vary over the primary beam, which is a major fraction of the sky. Even the frequency dependence of these effects is not simple at 1.5 MHz (equation 4). In general, the correction will not be a separable function of direction, frequency and polarization. Applying the corrections in the process of making an all-sky map will be a major computational problem.

The empirical data required for correction could come from in situ monitoring; this would yield the local medium delay in any direction with high accuracy. The global effects could be computed on the basis of a model. Since most of the refraction occurs in the lowest parts of the medium traversed, i.e. relatively close to the array, the model might be extrapolated from in-situ measurements with moderate accuracy. In situ monitoring may also be required to determine the system gain due to fluctuating antenna capacitance caused by variations in the plasma density. (Indeed, measurement of antenna capacitance could be used to determine the local plasma density.) A somewhat different approach to correcting the fringe shifts might involve monitoring the fringe phase of a known strong source.

IV. CONCLUSIONS

Our principal conclusion is that large orbital radii will be required for imaging at 1.5 MHz. The minimum radius has not been determined, but it is at least $2.5R_\oplus$ under conditions of solar maximum. Successful imaging from *within the plasmasphere* may depend upon the feasibility of correction schemes.

Clearly, much more extensive calculations are required, particularly for large radii, both propagation modes, and over a wide range of zenith angles. The magnetosphere should be modelled under both solar minimum and solar maximum conditions, and with a plasmapause and magnetotail. In addition, various possible correction schemes need to be considered in detail, and possibly simulated.

Finally, we note that considerations such as these need to be applied to the solar wind. It is of course recognized that scattering by irregularities will broaden compact sources and ultimately limit the resolution (Dennison and Booth 1987; Spangler 1990). In addition however, large-scale interplanetary refraction and the resulting distortion needs to be evaluated.

REFERENCES

Chui, Y. T., Luhmann, J. G., Ching, B. K., and Boucher, D. J. 1979, *J. Geophys. Res.*, **84**, 909.

Dennison, B. K., Weiler, K. W., Johnston, K. J., Simon, R. S., Spencer, J. H., Hammarstrom, L. M., Wilhelm, P. G., Erickson, W. C., Kaiser, M. L., Desch, M. D., Fainberg, J., Brown, L. W., and Stone, R. G. 1986, *Naval Research Laboratory Memorandum Report 5905*.

Dennison, B. K., and Booth, R. S. 1987, *M.N.R.A.S.*, **224**, 927.

Kasha, M. A. 1967, *The Ionosphere and its Interaction with Satellites*, (Gordon & Breach: New York).

Lang, K. R. 1980, *Astrophysical Formulae*, (Springer-Verlag: New York).

NASA 1971, *The Earth's Ionosphere*, SP 8049.

Spangler, S. R. 1990, this volume.

Weiler, K. W., Dennison, B. K., Johnston, K. J., Simon, R. S., Erickson, W. C., Kaiser, M. L., Cane, H. V., Desch, M. D., and Hammarstrom, L. M. 1988, *Astron. Astrophys.*, **195**, 372.

Weiler, K. W. et al. 1990, this volume.

MONITORING JUPITER'S HECTOMETRIC EMISSION

Thomas D. Carr and Liyun Wang
Department of Astronomy, University of Florida
Gainesville, Florida 32611

When it becomes possible to make observations of Jupiter's
hectometric radiation component (nominal range: 0.3 to 3 MHz) on a
regular basis, its importance will probably exceed that of the decametric radiation in producing new information relating to the magnetospheric environment and the internal field-generating dynamo mechanism
of the planet. We present the case in this paper for the long-term
monitoring of the Jovian hectometric radiation from a high-orbit
satellite or from the surface of the moon. Under the best terrestrial
observing conditions, i.e., between local midnight and dawn when solar
activity is near its minimum, ionospheric effects generally limit
regular observations of Jupiter's radio emissions from Earth-based
observatories to frequencies above about 7 MHz. Although Jovian
emission down to about 1 MHz has been detected from Earth orbit by
RAE 1, IMP 6, and ISEE 3, most of our knowledge of the hectometric
radiation has come from the data provided by Voyagers 1 and 2 [1,2,3,
4]. Unlike the decametric and kilometric components, which occur
sporadically, the hectometric radiation never fails to appear during
a Jovian rotation.

When a plot of intensity as a function of time from a receiving
channel in the vicinity of 1 MHz is suitably smoothed, there is a
strong tendency for a characteristic signature that is either bilobed
or monolobed to occur with each rotation of the planet. This is
illustrated in Figure 1, in which there are separate plots for both
Voyagers, both before and after Jovian encounter. Each plot spans
four consecutive rotations. The scales indicate day of the year 1979
(DOY), central meridian longitude (CML), and subsolar longitude (SSL).
It is apparent that the structure depends on CML, not on SSL. Such
plots at all frequencies from about 0.7 to 1.3 MHz were very similar,
but the repeating structure is best displayed at about 1 MHz.

In Figure 1, the pattern for each rotation is monolobed for the
plot of Voyager 2 before encounter, and is predominantly bilobed for
the other three plots. A total of 180 rotational cycles of data were
examined. For the post-encounter cycles observed from both spacecraft

Figure 1:

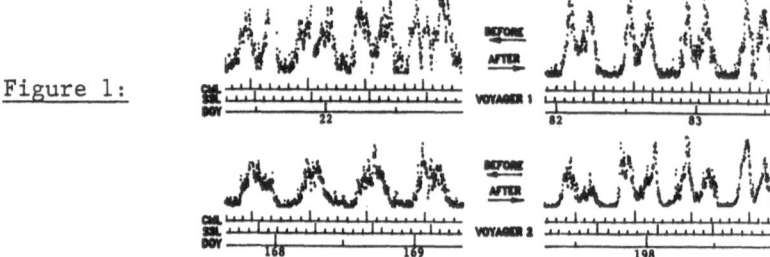

(at Jovicentric declinations of about 5°) the pattern was more or
less bilobed for about 75% of the rotations and monolobed for about
20%. The pattern for the pre-encounter data from Voyager 2 (declination 7°) was nearly always monolobed as in Figure 1, but was
occasionally bilobed. The predominance of monolobed patterns in this
case may be a consequence of the higher declination. For Voyager 1
before encounter (declination 3°) the patterns were irregular for
65%, recognizably bilobed for 25%, and monolobed for 10% of the
rotations, approximately. The increased irregularity in this case
we attribute to the location of the spacecraft closest to the ecliptic
plane and on the sunward side of Jupiter, where solar wind interactions
with the magnetosphere might interfere with the regularity of the
emission pattern most strongly. Other manifestations of a correlation
of Jovian hectometric and solar wind activity have been reported in
a number of papers (see reference 3, and references therein).

In an early paper, Alexander et al. [1] showed (using data of
relatively poor quality) that the variation of the occurrence probability of hectometric radiation with CML was roughly what would be
expected from a continuously emitted beam in the form of a thin
curved sheet at a fixed magnetic latitude that wobbles as the planet
rotates. More recently, wobbling thin-sheet beam models have been
considered by Carr and Wang (workshop paper presented at the Max
Planck Institut fur Aeronomie, Oct. 1988; also preprint, April 1989)
and by Ladreiter and Leblanc [3,4]. Both groups assumed cyclotron
maser emission from within particular L shells, producing the
rotation-modulated intensity patterns observed at the spacecraft.
According to the Carr-Wang model, the wide gaps are produced during
the part of the rotational phase at which the curved-sheet beam is
well south of the spacecraft, and the narrow gaps in the bilobed
patterns when the beam is slightly north of the spacecraft. For the
monolobed patterns, we assume that the declination of the spacecraft
is sufficiently high (northward) that the beam maximum direction never
passes north of it, so that there is only a single observed intensity
maximum during each rotation (i.e.,no narrow gap).

The midpoints of the wide and narrow gaps can be used as CML fiducial points to be measured and employed for rotation period determinations. Such a determination would involve the measurement of the drift in the CMLs of the gap midpoints that accumulates over a number of years because of the error in the assumed rotation period, and then correcting the asumed period so that the CML drift is eliminated. For the intensity vs CML curves obtained during the 180 rotational cycles that were examined, we have measured the CML values at the midpoints of the wide gaps and also at the midpoints of the narrow gaps when they were present. The mean values are presented in Table 1. The indicated errors are the standard deviations of the means. N is the number of measurements in each category, and SSL is the subsolar longitude of the gap midpoint (note the dependence on CML rather than SSL). The best estimate of the longitude of the north magnetic pole of offset tilted dipole models of the Jovian field based on in situ magnetometer measurements [5] is also given. The wide gap is symmetrical about the meridian of this pole to a remarkably high degree of accuracy.

Table 1. Measurement of Wide and Narrow Gap Midpoint Longitudes.

	Wide Gap			Narrow Gap		
	CML(°)	N	SSL(°)	CML(°)	N	SSL(°)
Voy.1 before enc.	203.8 ±1.9	47	180	31.0 ±2.0	32	7
Voy.2 " "	199.1 ±1.7	48	166	30.5 ±4.1	12	1
Voy.1 after enc.	203.0 ±1.4	74	87	30.4 ±1.8	59	275
Voy.2 " "	201.6 ±2.8	11	56	33.2 ±1.7	8	249
Weighted Mean	202.0 ±0.9			31.6 ±1.0		
N. Mag. Pole Long.	202.5					

It is known that the terrestrial magnetic field pattern is at present gradually moving westward, and that both the rate of drift and and the field configuration itself are highly variable over geological time. It has been suggested that such secular changes in the Jovian field may occur on much shorter time scales than those on Earth, and that they might at some not too distant time be detectable from radio measurements. The development of a dynamo theory of the Jovian magnetic field cannot proceed very far without observational data on the extent and nature of the secular variations. A continuing series of sufficiently precise magnetospheric rotation period measurements would eventually reveal the presence of magnetic polar wandering and would indicate its rate and variability. The only precise determinations that have thus far been made of Jupiter's rotation period were obtained from terrestrial measurements of its magnetospheric radio emissions, i.e., the

synchrotron emission at decimeter and centimeter wavelengths and the
decametric burst emission in the vicinity of 20 MHz. At present, it is
necessary to average the period over intervals of at least a decade in
order to obtain precisions of about 0.1 sec. The best synchrotron and
decametric determinations that have been made to date agree to within
0.05 sec [6]. Such measurements are not yet sufficiently precise for
the detection of a true change in the rotation period over the time span
that has been available. We believe that the high degree of consistency
of the measurements of the wide gap and narrow gap CML fiducial points
in Table 1 is an indication that the long-term monitoring of the hectometric radiation can provide rotation period data of higher precision
than can the present methods. If Galileo permits the measurement of
Jupiter's hectometric radiation with a precision comparable to that
with Voyager, it will be possible to use the combined Voyager and
Galileo data to measure the mean rotation period over the interval of
about 16 years with an accuracy of 0.01 sec or somewhat better. This
is about an order of magnitude better than could be done from terrestrial observatories over a comparable interval using the synchrotron
or decametric radiation.

We therefore recommend that long-term monitoring of the Jovian
radiation at a frequency of about 1 MHz be included in the planning
for observations to be made from high Earth orbit or from the lunar
surface. Terrestrial interference will not usually be a problem at
1 MHz, because the lightning transients from thunderstorms and the
radio station and other manmade interference will most often be completely blocked by the ionosphere, while the auroral kilometric radiation (AKR) rarely reaches frequencies as high as 1 MHz. If the observations are made from a spacecraft, the orbit should be sufficiently
high that Jupiter is unocculted by Earth for intervals of at least 10
hours. A sensitivity calculation indicates that with a receiver having
a bandwidth of 4 kHz and a post-detector time constant of 60 sec,
together with a short dipole antenna used as a voltage probe, the minimum detectable flux density would be about 50 kJy (50,000 Jy). This is
two orders of magnitude below the expected 24-hr peak value, and would
therefore be adequate. We propose that the spacecraft carry three
orthogonal short dipole antennas, each connected to a receiver with an
audio frequency baseband output; these signals would be taped and
later relayed to ground for further processing. In addition to providing sufficient information to aid in the recognition of interference
arriving from non-Jupiter directions, such a system would make possible
the determination of the Stokes parameters specifying the polarization
of the Jovian radiation.

Advantages of making the observations from the surface of the moon instead of from a satellite would be a) a phase-steerable dipole array could be used, providing higher sensitivity and better rejection of interference from non-Jupiter directions, b) continuous observations could be made for large fractions of each lunar month when the solar elongation is sufficiently high, and c) terrestrial interference, largely blocked by the ionosphere, would be further reduced because of the increased distance.

In summary, we recomend the long-term monitoring of Jupiter at a frequency of about 1 MHz either from high Earth orbit or from the lunar surface. Such a program would eventually provide a record of rotation period changes revealing secular variations in the Jovian magnetic field, with a sensitivity higher than could be achieved from terrestrial observatories. Such information is necessary for the development of a dynamo model of the Jovian field. On the shorter term, the program would provide a wealth of new data that would greatly facilitate the development of a complete theory of the origin of the Jovian hectometric radiation and its relationship to the solar wind, and would in addition provide a means for continuously monitoring an effect of the interaction of the solar wind with Jupiter's magnetosphere.

This work received support from NASA through Goddard Space Flight Center grant NAG 5-773 and Radiophysics, Inc. (Voyager PRA funds), and from National Science Foundation grant AST 8613453.

REFERENCES

1. Alexander, J.K., M.D. Desch, M.L. Kaiser, and J.L. Thieman, J. Geophys. Res., 84, 5167, 1979.
2. Alexander, J.K., T.D. Carr, J.R. Thieman, J.J. Schauble, and A.C. Riddle. J. Geophys. Res., 86, 8529, 1981.
3. Ladreiter, H.P., and Y. Leblanc. Astron. Astrophys., 226, 297, 1989.
4. Ladreiter, H.P., and Y. Leblanc. J. Geophys. Res., in press.
5. Acuna, M.H., K.W. Behannon, and J.E.P. Connerney. In Physics of the Jovian Magnetosphere, ed. A.J. Dessler, p. 1, Cambridge Univ. Press, New York, 1983.
6. May, J., T.D. Carr, and M.D. Desch. Icarus, 40, 87, 1979.

IV. SNe, SNRs, AND IONIZED GAS IN THE INTERSTELLAR MEDIUM

THE LOW DENSITY IONIZED COMPONENT OF THE INTERSTELLAR MEDIUM
AND FREE-FREE ABSORPTION AT HIGH GALACTIC LATITUDES

R. J. Reynolds
Department of Physics, University of Wisconsin-Madison
1150 University Avenue, Madison, WI 53706

ABSTRACT

At frequencies below 2 MHz the sky is opaque because of free-free absorption by warm, ionized interstellar gas. Diffuse H^+ is now known to be a major component of the interstellar medium, having a surface density approximately one third that of the H I, a scale height ten times that of the cloud layer, and a power requirement comparable to that available from supernovae. While the origin of this gas is not yet understood, its existence clearly has an important bearing on the nature of the interstellar medium and lower Galactic halo. Faint line emission from the gas at optical wavelengths and its free-free absorption pattern against the Galactic synchrotron background at very low radio frequencies provide opportunities to explore in detail the distributions of both this ionized component of the interstellar medium and the synchrotron emissivity within the Galactic disk and halo.

1. INTRODUCTION

In 1963 as a result of their analysis of the unique, very low frequency observations of the Galaxy obtained with the Reber radio array in Tasmania, Hoyle and Ellis [1] proposed the existence of an extensive layer of warm, ionized gas about the Galactic plane. Their recognition that the spatial and spectral characteristics of the Galactic synchrotron emission at frequencies below about 5 MHz were dominated by free-free absorption led them to derive a mean emission measure $(EM = \int_{-\infty}^{\infty} n_e^2 \, dz)$ of 5 cm^{-6} pc and a free electron column density $(N_{H^+} = \int_{-\infty}^{\infty} n_e \, dz)$ of order 10^{20} cm^{-2} along a line perpendicular to the Galactic disk. They also concluded that the total radiative power from this ionized gas approached the total ionizing power of the O and B stars. Unfortunately, this remarkable paper was essentially ignored by Galactic and interstellar astronomers. Years later the discoveries of pulsar dispersion measures and diffuse Galactic $H\alpha$ emission revealed again the existence of a widespread interstellar ionization having properties nearly identical to those derived by Hoyle and Ellis. The f-f absorption by this gas at very low frequencies also was confirmed by Radio Astronomy Explorers 1 and 2 [2, 3].

A close relationship thus exists between pulsar dispersion measures, diffuse optical line emission, and Galactic radio emission (more accurately, absorption) at frequencies below 5 MHz. The following is a brief review of the properties of the diffuse ionized gas derived from the pulsar and optical line data and its implications for observations of Galactic and extragalactic emission at very low frequencies.

2. THE DIFFUSE INTERSTELLAR H^+

2.1. Properties of the Ionized Gas

The pulsar dispersion data directly confirm the existence of a layer of ionized hydrogen about the Galactic plane having a thickness $2H \approx 2$ kpc and a space averaged electron density $\langle n_e \rangle_0 \approx 0.03$ cm^{-3} near the midplane [4, 5]. This layer has an H^+ surface density $N_{H^+} \approx 2 \times 10^{20}$ cm^{-2}, which is about one-third the H I surface density in the solar neighborhood. Table 1 compares H^+ with H I column densities toward the five globular cluster pulsars that have distances above the midplane $|z| > 3$ kpc, which place them above more than 90% of the H^+. These data show that the H^+ is a significant component of the interstellar medium, particularly along lines of sight at high Galactic latitude. The minimum power required to maintain this ionization, $P \approx 1 \times 10^{-4}$ ergs s^{-1} per cm^2 of Galactic disk [13], is comparable to the total power injected into the diffuse interstellar medium by luminous stars and supernovae. This power requirement also suggests that most of this gas, even at high $|z|$, is at a temperature near 10^4 K, since gas temperatures that are much higher or lower would imply more power than is available from known sources [13].

Table 1

COMPARISONS OF H^+ and H I COLUMN DENSITIES AT HIGH GALACTIC LATITUDE[*]

l	b	N_{H^+} (10^{20} cm^{-2})	N_{HI} (10^{20} cm^{-2})	N_{H^+}/N_{HI}
4°	+47°	0.91	4.0	0.23
59	+41	0.94	1.5	0.63
65	−27	2.1	6.2	0.34
306	−45	0.77	2.75	0.28
333	+80	0.74	2.0	0.37

*Toward globular cluster pulsars with $|z|$ distances > 3 kpc, which place them above > 90% of the H^+. Values for N_{H^+} are from Anderson et al. [6,7,8], Wolszczan et al. [9], and Manchester et al. [10]. Values for N_{HI} are from Crawford Hill [11] and Parks [12] 21 cm line surveys.

Hydrogen recombination within this gas produces diffuse interstellar Hα emission, which has been studied with high throughput Fabry-Perot spectroscopy [14]. The intensity I_α of the Hα in a line of sight s that is not affected by interstellar extinction is given by

$$I_\alpha = \frac{1}{4\pi} \int \alpha_{H\alpha}(T) \; n_e^2 \; ds, \qquad\qquad (1)$$

where n_e is the electron density, $\alpha_{H\alpha} = 1.17 \times 10^{-13} \; T_4^{-0.92}$ Hα photons $cm^3 \; s^{-1}$ [15], and T_4 is the temperature in units of 10^4 K. Thus observations of the Hα intensity directly provide values for $EM \cdot T^{-0.92}$ within the Galactic disk at latitudes $|b| \gtrsim 5°-10°$, where interstellar extinction of the Hα can be neglected. Fabry-Perot scans have sampled the interstellar Hα in a few hundred directions on the sky, most with a 0.°8 diameter beam. From these data a crude map of the Hα intensity distribution has been synthesized [14]. The diffuse Galactic Hα appears to cover the entire sky with an intensity that ranges from about 5×10^5 photons $cm^{-2} \; s^{-1} \; sr^{-1}$ near the Galactic equator to about 4×10^4 photons $cm^{-2} \; s^{-1} \; sr^{-1}$ near the Galactic poles. At $|b| \gtrsim 10°$ the intensities generally follow the $csc|b|$ law. The mean value for $I_\alpha \cdot \sin |b|$ at high latitudes implies $\langle EM \cdot T_4^{-0.92} \rangle = 5.5 \; cm^{-6}$ pc along a line perpendicular to the disk.

In order to investigate the Hα intensity distribution on smaller angular scales, two $11° \times 11°$ regions of the sky were mapped on an approximately $1° \times 1°$ grid with a 0.°8 diameter beam [16, 17]. One of these maps was centered at $l = 144°$, $b = -20°$ and the other at $l = 213°$, $b = +3°$. The resulting $1°$ angular resolution, 12 km s^{-1} velocity-interval "images" of the Hα background reveal a complex morphology consisting of filaments and diffuse patches of enhanced emission that are superposed on a fainter, more uniform Hα background (see, for example Fig. 2 below). The nature of these enhancements as well as the more diffuse background remains unknown. These observations clearly illustrate the tremendous amount of new information about interstellar matter and processes that is present in the Hα background. An examination of the structure of the background emission at even higher angular resolution (~ 1'-2') has been started recently by directly imaging the 0.°8 diameter sky fields onto a low noise CCD [18, 19]. Preliminary results from this work suggest that the emission is relatively smooth on angular scales ~ 0.°1, at least at moderate to high Galactic latitudes.

A comparison of Hα intensities, which probe $\int n_e^2 \; ds$, with pulsar dispersion measures, which probe $\int n_e \; ds$, indicates that the ionized gas is clumped into regions that occupy about 20% of the volume with a density of 0.15 cm^{-3} near the Galactic midplane [13]. Such densities are consistent with the densities derived from the Hα intensity enhancements in the high (1°) angular resolution maps (e.g., [16]).

In addition to Hα the nebular lines [N II] λ6583, [S II] λ6716, and [O III] λ5007 have been identified in the interstellar background. This emission provides additional clues about the physical conditions within the emitting gas. An analysis of the line widths of Hα and [S II] provides an upper limit of 20,000 K on the gas temperature and a most probable temperature of about 8000 K [20]. This line width analysis also indicates that the random, nonthermal motions within the gas have speeds of 10-30 km s^{-1}. The existence of the thermally excited forbidden line emission also provides a lower limit on the electron temperature of about 5500 K [21]. Hereafter a temperature of 8000 K is adopted for the emitting gas. Searched for but not found are [N I] λ5201 and [O I] λ6300. Their absence implies that within the ionized regions the hydrogen is nearly fully ionized, having an ionization fraction $n(H^+)/n(H^0) > 2$ [21].

A summary of the general properties of this ionized gas is presented in Table 2.

2.2. Significance of the H$^+$ to Galactic Astronomy

The interstellar H$^+$ is the only major component of the interstellar medium that has not yet been surveyed. Neither the source of this widespread ionization nor its relationship to the other components of the interstellar medium is

Table 2

GENERAL CHARACTERISTICS OF THE WARM, IONIZED

COMPONENT OF THE INTERSTELLAR MEDIUM

Surface Density..............$N_{H^+} \simeq 2 \times 10^{20}$ cm^{-2}

Emission Measure.............EM $\simeq 4$ cm^{-6} pc

Scale Height.................H $\simeq 1$ kpc

Midplane Density.............$\langle n_e \rangle_0 \simeq 0.025$ cm^{-3}

Filling Fraction.............f $\simeq 0.2$

Temperature..................T $\simeq 8000$ K

Pressure.....................$P_0/k \simeq 3000$ cm^{-3} K

Ionization Fraction..........$H^+/H^0 \gtrsim 2$

Power Consumption............W $\simeq 1 \times 10^{-4}$ ergs s^{-1} cm^{-2}

understood. In the standard McKee and Ostriker [22] model of the interstellar medium it was suggested that the warm, ionized gas is located in transition regions between the hot, "coronal" medium and the H I clouds, and that it is ionized by ambient radiation from O stars and supernova remnants. However, this picture appears to be incompatible with the fact that most of the H^+ is located well above the layer of stars and H I clouds that define the traditional disk of the Galaxy. Thus the existence of this gas has an important bearing on our understanding of the composition of the interstellar medium and lower Galactic halo and the principal processes of heating and ionization within them.

In addition, the existence of the diffuse H^+ has a significant impact on the interpretation of observations obtained at radio, IR, UV, EUV, and even γ-ray wavelengths [14, 23]. For example, the ionized gas may account on average for about 20% of the observed Galactic background emissions in the far-UV (2-γ emission of hydrogen) and IR (100 μm emission from associated warm dust). The ionized hydrogen also contributes to the diffuse γ-ray background produced by the interaction of high energy cosmic rays with interstellar matter. As a result the H^+ distribution will affect the derived value for the cosmic ray flux within the Galactic disk, especially at high Galactic latitudes, where in some directions the column density of the $H^+ \gtrsim 50\%$ of the H I (see Table 1; also [23]). The H^+ also impacts UV absorption line studies (e.g., the origin of moderate to low ions such as S^+), observations in the EUV (He I absorption within the H^+ regions), and, of course, radio continuum observations (e.g., rotation measures; f-f absorption at very low frequencies).

3. FREE-FREE ABSORPTION BY THE WARM, IONIZED MEDIUM

The f-f optical depth at frequency ν through an ionized region of the interstellar medium is given by

$$\tau_{ff} = 0.53 \ \nu_{MHz}^{-2} \ T_4^{-1.5} \ g(\nu,T) \ EM, \qquad (2)$$

where EM is in units of cm^{-6} pc and $g(\nu,T) \simeq 1 + 0.085 \ln T_4 - 0.056 \ln \nu_{MHz}$ [24]. At low Galactic latitudes (i.e., $|b| < 5°$), where traditional H II regions having $EM \gtrsim 10^3 \ cm^{-6}$ pc are intercepted by the lines of sight, the Galactic disk should be optically thick at $\nu \lesssim 30$ MHz. On average these traditional H II regions will provide an emission measure of 200 cm^{-6} pc per kpc, which is much larger than the emission measure of 5-10 cm^{-6} pc per kpc associated with the widespread, intercloud ionization discussed in Section 2 above.

On the other hand, at $|b| \gtrsim 5°$ very few H II regions are present, and the f-f optical depth along the lines of sight will be determined solely by the intercloud ionization. The values of $\tau_{ff}(\nu)$ at high latitude can be estimated directly from the intensity of the interstellar Hα background. From equations (1) and (2) the frequency at which the f-f optical depth is unity is given by

$$\nu(\tau_{ff} = 1) \approx 1.2 \; T_4^{-1/4} \; I_\alpha^{1/2} \; \text{MHz}, \qquad (3)$$

where I_α is in rayleighs R; $1 \; R = 10^6/4\pi$ photons $\text{cm}^{-2} \; \text{s}^{-1} \; \text{sr}^{-1}$.

Figure 1 is a map of predicted values for $\nu(\tau_{ff} = 1)$ over the sky obtained from equation (3) and the low angular resolution Hα map [14]. The Hα observations used to create this map are scattered over the sky at declinations greater than -20° with the angular separations between the observation directions typically ranging from 5°-10° at $|b| \lesssim 20°$ to 10°-30° at higher latitudes. This extreme undersampling of the sky obviously restricts the information that can be obtained from this map primarily to the very large scale features of the diffuse ionization. The values of $\nu(\tau_{ff} = 1)$ range from about 1 MHz near the Galactic poles to 4 MHz at $|b| \approx 5°$. Variations in $\nu(\tau_{ff} = 1)$ across the sky at a given Galactic latitude appear to be approximately a factor of two (peak to valley).

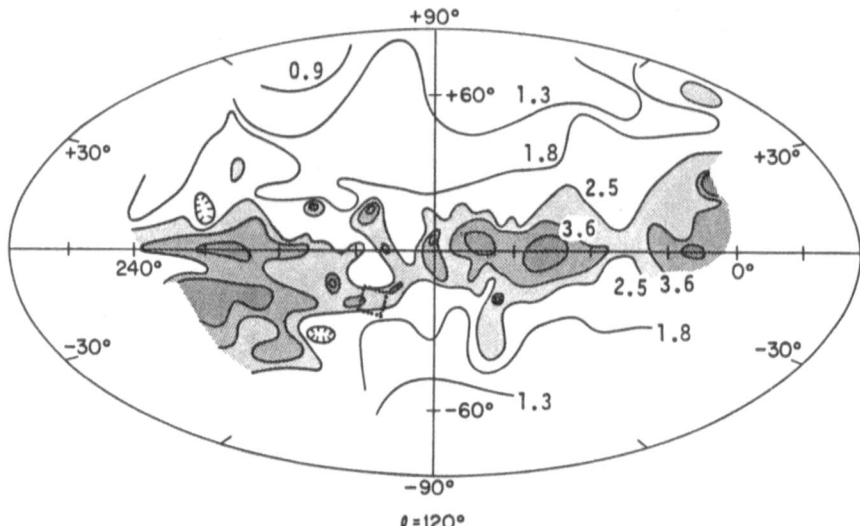

Fig. 1 - A map of the sky (centered at l = 120°, b = 0°) showing the frequency at which τ_{ff} = 1. Contour values were derived from the observed intensity of interstellar Hα [14] and eq. (3). The information in this map is limited to the very large angular scale ($\gtrsim$ 10°-30°) features of the free-free absorbing medium (see text). The dotted rectangle denotes the 12° x 11° region in which information at higher angular resolution ($\approx$ 1°) is available (see Fig. 2 and Table 3). At $|b| \lesssim 5°$ the contour values provide only a lower limit to $\nu(\tau_{ff}$ = 1) because of significant interstellar extinction of the Hα.

The high angular and spectral resolution Hα observations that have been carried out for two small areas of the sky [16, 17] show that there is much more information available from the Hα background then that presented in Figure 1. These higher resolution maps reveal a complex pattern of Hα intensity enhancements having angular sizes $\gtrsim 2°$ and emission measures of $1-10$ cm^{-6} pc. The radial velocities of these regions of enhanced emission are due in part to Galactic differential rotation, which can provide kinematic distances and thus locate these regions within the Galactic disk. Figure 2 shows five such regions identified within a $12° \times 11°$ region of the sky centered near $l = 144°$, $b = -20°$ (dashed rectangle in Fig. 1). For each of these regions the Hα intensity, emission measure, kinematic distance, height above the midplane, and value of $\nu(\tau_{ff} = 1)$ are listed in Table 3. Note that some of the H$^+$ regions extend more than 1 kpc from the midplane into the lower Galactic halo. These high $|z|$ regions will thus selectively absorb only low frequency emission originating from the outer Galactic halo and extragalactic sources.

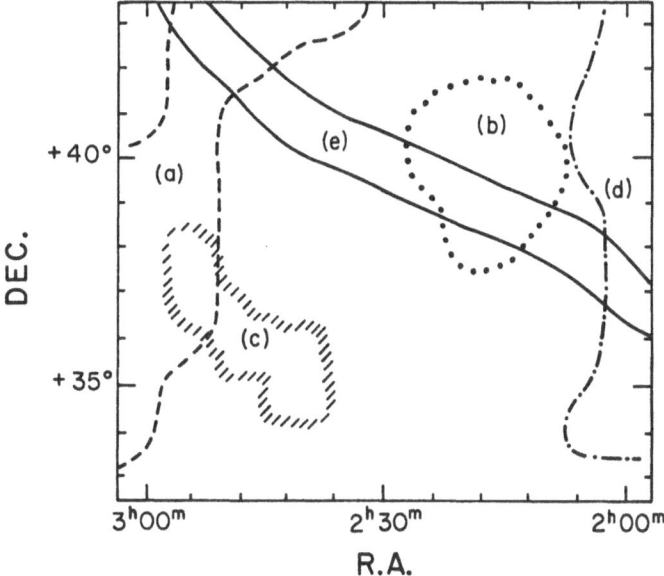

Fig. 2 - Five "clouds" of ionized gas within a $12° \times 11°$ region of the sky centered near $l = 144°$, $b = -20°$ (dotted rectangle in Fig. 1). These clouds are visible as regions of Hα intensity enhancement on radial velocity interval maps having approximately 1° angular resolution and 12 km s^{-1} velocity resolution (see [16]). Table 3 below lists for each region its average Hα intensity above the more diffuse background, emission measure (for $T_4 = 0.8$), kinematic distance, z-height from the Galactic midplane, and the frequency at which the cloud becomes optically thick due to f-f absorption (from eq. 3).

Table 3

PREDICTED FREQUENCIES FOR $\tau_{ff} = 1$ IN LOW DENSITY H^+

REGIONS AT MODERATE GALACTIC LATITUDE (FIG. 2)

region	I_α (R)	EM (cm^{-6} pc)	d (kpc)	$|z|$ (kpc)	$\nu(\tau_{ff} = 1)$ (MHz)
a	5.3	12	< 1.2	< 0.4	2.9
b	1.2	2.7	1.6	0.5	1.4
c	0.8	1.8	3.1	1.1	1.1
d	1.0	2.3	3.5	1.3	1.3
e	0.8	1.8	5.4*	1.8*	1.1

*A kinematic distance may not be appropriate for this region (see [25]).

4. CONCLUSIONS

Figures 1 and 2 and Table 3 show that Galactic and extragalactic observations at frequencies between 1 and 4 MHz will be greatly influenced by the warm, ionized component of the interstellar medium. For example, the distribution of the ionized gas will produce a complex f-f absorption shadow pattern against the sky at low frequencies. Thus a radio continuum sky survey of the Galactic synchrotron background at frequencies down to 1 MHz would provide valuable information about the distribution of this major, but little understood ionized component of the interstellar medium. Furthermore, a comparison of the free-free shadow pattern on the synchrotron background with $H\alpha$ survey data, which can give both the τ_{ff} and the locations (through Galactic rotation) of the absorbing H^+ regions, would provide information about the distribution of the synchrotron emission within the disk and halo.

The technology is now available to build a high throughput Fabry-Perot facility capable of mapping at moderate (1°) angular resolution the faint, Galactic $H\alpha$, [S II], and [N II] emission-line background associated with the warm (10^4 K), ionized component of the interstellar medium [26]. Hopefully, this ground-based, optical survey can be carried out within the next few years. This survey plus a space-based radio survey at comparable angular resolution of the Galactic continuum background at frequencies down to 1 MHz could provide important, entirely new perspectives on the composition and structure of the interstellar medium and the distribution of synchrotron emissivity within the Galaxy.

This research has been supported by the National Science Foundation through grants AST88-13467 and AST86-10431.

REFERENCES

1. F. Hoyle, and G. R. A. Ellis, Australian J. Phys., 16, 1, 1963.
2. J. K. Alexander, L. W. Brown, T. A. Clark, and R. G. Stone, Astron. Astrophys., 6, 476, 1970.
3. J. C. Novaco, and L. W. Brown, Astrophys. J., 221, 114, 1978.
4. J. M. Weisberg, J. M. Rankin, and V. Boriakoff, Astron. Astrophys., 88, 84, 1980.
5. R. J. Reynolds, Astrophys. J. (Lett.), 339, L29, 1989.
6. S. Anderson, P. Gorham, S. Kulkarni, T. Prince, and A. Wolszczan, IAU Circular No. 4772, 1989.
7. S. Anderson, S. Kulkarni, T. Prince, and A. Wolszczan, IAU Circular No. 4819, 1989.
8. S. Anderson, S. Kulkarni, T. Prince, and A. Wolszczan, IAU Circular No. 4853, 1989.
9. A. Wolszczan, S. Anderson, S. Kulkarni, and T. Prince, IAU Circular No. 4880, 1989.
10. R. N. Manchester, A. G. Lyne, S. Johnston, N. D'Amico, J. Lim, and D. A. Kniffen, IAU Circular No. 4892, 1989.
11. A. A. Stark, C. Gammie, J. Bally, R. A. Linke, C. Heiles, and R. W. Wilson, 1989, in preparation.
12. D. McCammon, S. S. Meyer, W. T. Sanders, and F. O. Williamson, Astrophys. J., 209, 46, 1976.
13. R. J. Reynolds, Astrophys. J. (Lett.), 349, L17, 1990.
14. R. J. Reynolds, in Galactic and Extragalactic Background Radiation, IAU Symposium No. 139, ed. S. Bowyer and C. Leinert (Dordrecht: Kluwer), 1989, in press.
15. P. G. Martin, Astrophys. J. Suppl., 66, 125, 1988.
16. R. J. Reynolds, Astrophys. J., 236, 153, 1980.
17. R. J. Reynolds, Astrophys. J., 323, 118, 1987.
18. S. Milster, Ph.D. thesis, UW-Madison, in progress, 1990.
19. J. Brinkmann, Ph.D. thesis, UW-Madison, 1987.
20. R. J. Reynolds, Astrophys. J., 294, 256, 1985.
21. R. J. Reynolds, Astrophys. J., 345, 811, 1989.
22. C. K. McKee, and J. P. Ostriker, Astrophys. J., 218, 148, 1977.
23. H. Bloemen, Annu. Rev. Astron. Astrophys., 27, 469, 1989.
24. R. Gayet, Astron. Astrophys., 9, 312, 1970.
25. P. M. Ogden, and R. J. Reynolds, Astrophys. J., 290, 238, 1985.
26. R. J. Reynolds, F. L. Roesler, F. Scherb, and J. Harlander, in Instrumentation in Astronomy VII, SPIE Proceedings No. 1235, 1990.

SUPERNOVAE AND SUPERNOVA REMNANTS AT LOW FREQUENCIES

Roger A. Chevalier
Department of Astronomy, University of Virginia
P. O. Box 3818, Charlottesville, VA 22903

I. Introduction

The radio emission from supernova remnants falls into two basic types: shell remnants with a relatively steep radio spectrum and filled-center (or plerion) remnants with a relatively flat spectrum. Some remnants show both types of emission (e.g. Weiler and Sramek 1988). In both cases, the emission is thought to be synchrotron radiation from relativistic electrons.

The theory for the production of relativistic electrons and the amplification of magnetic fields is not on a sound basis. In a remnant like Cas A, there are random motions of knots suggesting that some kind of turbulent acceleration is taking place. Models of this type have been developed for Cas A (Gull 1973; Chevalier, Oegerle, and Scott 1978; Cowsik and Sarkar 1984). While these models can reproduce the behavior of the Cas A radio emission, they are not entirely convincing because there are a number of poorly constrained parameters. A more widely accepted mechanism for the acceleration of relativistic particles is first-order Fermi acceleration in a shock front (e.g. Blandford and Eichler 1987 and references therein). For a strong shock wave with a factor 4 density compression, a power-law energy spectrum of the form E^{-2} is naturally produced. While this spectral index is close to that inferred in young supernova remnants, the observations show considerable dispersion in the spectral index. This dispersion is not understood. However, the very sharp turn-on of the radio emission in remnants like Tycho and SN1006 suggests that some form of shock acceleration is taking place.

In older supernova remnants, there is an excellent correlation between optical filaments and the radio emission structure. The strong compression behind a radiative shock front presumably gives rise to the enhanced radio emission; this is the van der Laan (1962) emission mechanism. The emission may be due to a combination of shock wave acceleration and the postshock compression.

While the radio emission from shell remnants is poorly understood, that from plerions is even more of a mystery. It is supposed that a broad energy spectrum of relativistic electrons can be produced at the shock front where the pressure of a relativistic pulsar wind drops to the pressure of the piled-up wind. Kennel and Coroniti (1984) have used such a model to explain the optical through γ-ray emission from the Crab Nebula. However, the model fails to explain the observed radio emission.

Young plerions may be present inside some Type II supernovae, but the strong free-free absorption by the supernova envelope is expected to prevent their detection for many years. The radio emission that has been observed from supernovae (Weiler et al. 1986) is plausibly interpreted as synchrotron emission from the interaction of the supernova with a circumstellar wind. The mechanism for the emission is probably similar to that in young supernova remnants.

The shell supernova remnants are especially important for low frequency observations because they tend to have steeper radio spectra. Also, the shell emission is much more frequently observed than the filled-center emission. It will be emphasized here. Section 2 contains a discussion of the intrinsic nonthermal emission from supernovae and their remnants. Absorption processes, especially free-free absorption, are covered in section 3. Prospects for low frequency observations are summarized in section 4.

2. Intrinsic Emission

A common assumption in the study of radio supernova remnants is that their spectra are power laws over the entire radio range. While current theoretical ideas do allow for the production of a power law energy spectrum, they do not account for the observed range of spectral indices. The possibility of spectral evolution and of variations of spectral index across the face of a supernova remnant cannot be ruled out. Here, we summarize the observational evidence on these matters.

Some basic properties of radio supernovae are given in Table 1. The first group contains objects that were probably observed within a few years of the explosion. The spectral index α is defined by flux $S_\nu \propto \nu^{-\alpha}$. The quantity t_{20cm} is an estimate of the age at which the optical depth to free-free absorption in the circumstellar medium is 1 at 20 cm. The observed radio spectra are generally heavily affected by absorption and the values of α are from model fits to the radio evolution (Weiler et al. 1986). At late times the sources do become optically thin, but it is not clear whether there is any evolution of the

Table 1 - Radio Supernovae

SN	Type	Ratio to Cas A to 6cm	Spectral index α	Ref.	t_{20cm} (days)
1970G	II	~ 15	0.7±0.1	a	
1979C	II	~ 250	0.72±0.05	a	950
1980K	II	~ 15	0.50±0.06	a	190
1981K		~ 15	0.91±0.07	a	
1983N	Ib	~ 125	1.03±0.06	a	30
1984L	Ib	~ 60	1.0±0.2	a	
1986J	II	~3000	0.67±0.06	b	1600
1987A	II	~0.02	1.0±0.1	c	2
1950B		~ 5	0.4	d	
1957D		~ 15	0.25	d	
1961V		~ 1	0.4±0.3	e	

a) Weiler et al. 1986

b) Weiler, Panagia, and Sramek 1990

c) Turtle et al. 1987

d) Cowan and Branch 1985

e) Cowan, Henry, and Branch 1988

intrinsic spectrum. There is a considerable range in the spectral indices of individual objects.

The second group of radio supernovae includes objects observed at an age of decades and it is not clear whether the situation is the same as for the younger ones. SN 1961V may not even be a supernova (Goodrich et al. 1989). For the other two, there is a chance that plerion emission is beginning to play a role; the sources do show flatter spectra.

Table 2 summarizes spectral index data on supernova remnants; the data are from compilations by Green (1984) and Weiler et al. (1986). The division between young shells and old shells is based on whether radiative shock waves are present (old) or not (young). The objects are chosen as the best known members of the classes. It is interesting that the emission associated with radiative shock fronts has a systematically flatter spectrum than that from the young shells.

The detailed spectral properties of an individual object are best exemplified by Cas A, which is the brightest nonthermal radio source outside the solar system and the best studied. It has been known for decades that the total flux is

Table 2 – Supernova Remnants

Young Shells (no radiative shocks)	α
SN 1006	0.6
SN 1572 (Tycho)	0.61
SN 1604 (Kepler)	0.64
Cas A	0.77
Old Shells (radiative shocks)	
Cygnus Loop	0.45
IC 443	0.4
S 147	0.5
Vela XYZ	0.3
Plerions (filled-center)	
3C 58	0.1
SN 1054 (Crab)	0.26

declining, in rough agreement with Shklovsky's prediction based on adiabatic expansion. However, the situation is more complicated. The spectrum also appears to be flattening with time (Baars and Hartsuijker 1972; Dent, Aller, and Olsen 1974; Baars et al. 1977). Baars et al. (1977) express the frequency-dependent decline as

$$d(\nu)[\% \text{ per yr}] = 0.97 \ (\pm 0.04) - 0.30 \ (\pm 0.04) \ \log \nu \ [\text{GHz}].$$

The spectral region 20-300 MHz is somewhat flatter than the 0.3-30 GHz region, perhaps due to absorption. Models have been able to reproduce the general features of the observed spectral evolution (Chevalier, Oegerle, and Scott 1978; Cowsik and Sarkar 1984), but the models are not well constrained.

Particularly remarkable is a "flare" that was observed from Cas A at 38 MHz in the mid-1970's (Erickson and Perley 1975; Read 1977a, b). The total flux evolved on a timescale of 2-6 years. When this timescale is combined with a knot velocity of 6000 km s^{-1}, the maximum linear size is 0.05 pc, or an angular size of about 4". At its peak, the "flare" flux was about 25% above the expected 38 MHz flux and the excess emission did not appear at 80 MHz and higher frequencies. The source of the time variable, compact, steep-spectrum emission is unknown. Low frequency interferometry would be valuable to study such features. Low frequency mapping would also be useful to search for spectral index variations across supernova remnants. Rosenberg (1970) presented a spectral index map of Cas A from the frequencies 1.4 to 2.7 GHz. The map shows variations in α of

±0.15 about the mean of 0.75, but the variations may be due to observational error. A larger frequency range would show up differences in α more easily.

Cas A appears to be a rich subject for study, but its complex, knotty morphology is unusual and other young remnants may not be as active. For example, Vinyaikin et al. (1987a) have studied Tycho's remnant over the frequency range 12.6 MHz to 15 GHz and find an excellent power law fit over the entire range with α = 0.61 ± 0.03.

3. Absorption Processes

A number of absorption processes can become important at low frequencies, including synchrotron self-absorption, the Razin-Tsytovich plasma effect, and free-free absorption. For the conditions present in supernovae and their remnants, free-free absorption by free electrons appears to be the dominant process.

When a Type II supernova occurs, an initial flash of ionizing radiation is expected. The interaction of the supernova with presupernova mass loss can also generate ionizing radiation. Free-free absorption by circumstellar matter outside the shock interaction region is probably the cause of the low frequency absorption observed in radio supernovae (Chevalier 1982; Weiler et al. 1986). As the shock progresses through the circumstellar matter, a process which might last for decades, the absorption optical depth is reduced. Table 1 gives the ages at which radio supernovae have reached optical depth unity at 20 cm. For supernovae in which the external medium is dense, like SN 1979C, SN 1980K, and SN 1986J, the supernova may remain optically thick in the low frequency range 1-30 MHz throughout the circumstellar interaction. However, for the lower density cases like the Type Ib supernovae and SN 1987A, the source should become optically thin at low frequencies. The absorption process can then be followed over a large range of frequencies or, equivalently, a large range of shock radii. Because the absorption is related to the density of the surrounding gas, these cases give the most information on the structure of the surrounding medium. SN 1987A was observed in the radio for the longest time at the lowest frequency, 842 MHz (Turtle et al. 1987).

The ionizing flash from a Type II supernova can also ionize the surrounding medium on a scale of parsecs. Such ionization might also be present if the supernova precursor is an early type star. Vinyaikin et al. (1987b) argue that an HII region around the supernova remnant Cas A is responsible for the low

frequency turnover observed in its spectrum. They note that diffuse Hα emission
in the general direction of Cas A implies an insufficient free-free optical depth
due to the interstellar medium. However, a patchy HII region in the immediate
vicinity of the remnant has an emission measure of 740 pc cm^{-6} (Peimbert 1971),
which can easily account for the observed optical depth. It is quite possible
that the progenitor of Cas A was a Wolf-Rayet star, which could have created the
ionized region. Ionized regions have also been observed around the remnants
N132D (Lasker 1978) and 1E 0102.2-7219 (Tuohy and Dopita 1983) in the Magellanic
Clouds.

Once supernova remnant shock waves enter the radiative phase of evolution,
another source of free-free absorption arises. The radio emission in this case
probably comes from the dense layer of gas that builds up behind the radiative
shock front. To reach the observer, the radio emission must pass through the
recombining cooling layer where the ionized gas densities can be quite high
because of the high gas pressure. From the shock models of Raymond (1979), we
estimate that the emission measure is about $10^2 (n_o/10\ cm^{-3})$pc cm^{-6}, where n_o is
the preshock density, for shock velocities in the commonly observed range (80-100
km s^{-1}). This emission measure applies for free-free absorption due to a face-on
shock front. The effect is increased if the shock front normal is at an angle to
the line of sight.

In addition to absorption effects local to the supernova remnants, there is
absorption along the line of sight through the interstellar medium. Kassim
(1989) has recently completed a study of supernova remnants at 30.9 and 57.5 MHz
in order to derive detailed low frequency spectra for 32 supernova remnants. He
finds evidence for absorption in 2/3 of the objects; the amount of absorption is
poorly correlated with the distance to the remnant, implying the presence of a
widespread, but inhomogeneous ionized absorbing medium. Kassim identifies the
absorption as probably being associated with extended HII region envelopes, first
seen in radio recombination line studies at Ooty (Anantharamaiah 1986). These
envelopes have $T_e \sim$ 3000 - 8000K, $n_e \approx$ 0.5 - 10 cm^{-3}, emission measures of 500-
3000 pc cm^{-6}, sizes of 50-200 pc, and a filling factor <1%. Kassim notes that
normal HII regions may also play a role. Here we note that some of the remnants
may be affected by absorption in their immediate vicinity, as discussed above.
This will become more likely as the low frequency absorption data is pushed to
lower frequencies.

4. Summary

The intrinsic radio emission from supernovae and their remnants is still poorly understood so that observations over the broadest possible frequency range can be very useful. An example of a peculiarity at low frequencies is the 38 MHz "flare" observed from Cas A in the mid-1970's. Such events could provide the key to understanding the radio emission from a remnant like Cas A.

The radio emission from supernovae is so strongly free-free absorbed by circumstellar gas that explosions in a dense wind may never be detectable at low frequencies. However, for explosions in a lower density wind it may be possible to follow the absorption over a large range of shock front radii. Free-free absorption can also affect the radio emission from supernova remnants. Absorption local to the remnants can occur in HII regions created by the supernovae or their progenitors and in the cooling layers of radiative shock fronts. Alternatively, absorption can occur in the ionized interstellar medium. Observations of emission as well as absorption are probably necessary to determine the location of the ionized gas.

The author's research on supernova remnants is partially supported by NASA Grant NAGW-764.

References

Anantharamaiah, K. R. 1986, J. Ap. Astr., 7, 131.
Baars, J. W. M. and Hartsuijker, A. P. 1972, Astr. Ap., 17, 172.
Baars, J. W. M., Genzel, R., Pauliny-Toth, I. I. K., and Witzel, A. 1977, Astr. Ap., 61, 99.
Blandford, R. D. and Eichler, D. 1987, Phys. Reports, 154, 2.
Chevalier, R. A. 1982, Ap. J., 259, 302.
Chevalier, R. A., Oegerle, W. R. and Scott, J. S. 1978, Ap. J., 222, 527.
Cowan, J. J., and Branch, D. 1985, Ap. J., 293, 400.
Cowan, J. J., Henry, R. B. C., and Branch, D. 1988, Ap. J., 329, 116.
Cowsik, R. and Sarkar, S. 1984, M.N.R.A.S., 207, 745.
Dent, W. A., Aller, H. D., and Olsen, E. T. 1974, Ap. J. (Letters), 188, L11.
Erickson, W. C. and Perley, R. A. 1975, Ap. J. (Letters), 200, L83.
Goodrich, R. W., Stringfellow, G. S., Penrod, G. D., and Filippenko, A. V. 1989, Ap. J., 342, 908.
Green, D. 1984, M.N.R.A.S., 209, 449.
Gull, S. F. 1973, M.N.R.A.S., 162, 135.
Kassim, N. E. 1989, Ap. J., 347, 915.
Kennel, C. F. and Coroniti, F. V. 1984, Ap. J., 283, 710.
Lasker, B. M. 1978, Ap. J., 223, 109.
Peimbert, M. 1971, Ap. J., 170, 261.
Raymond, J. C. 1979, Ap. J. Suppl., 39, 1.
Read, P. L. 1977a, M.N.R.A.S., 178, 259.
Read, P. L. 1977b, M.N.R.A.S., 181, 63[b].

Rosenberg, I. 1970, M.N.R.A.S., 151, 109.

Turtle, A. J., Campbell-Wilson, D., Bunton, J. D., Jauncey, D. L., Kesteven, M. J., Manchester, R. N., Norris, R. P., Storey, M. C., and Reynolds, J. R. 1987, Nature, 327, 38.

Tuohy, I. R. and Dopita, M. A. 1983, Ap. J. (Letters), 268, L11.

van der Laan, H. 1962, M.N.R.A.S., 124, 125.

Vinyaikin, E. N., Volodin, Yu.V., Dagkesamanskii, R. D., and Sokolov, K. P. 1987a, Sov. Astr. AJ, 31, 141.

Vinyaikin, E. N., Nikonov, V. A., Tarasov, A. F., Tokavev, Yu.V., and Yurishchev, M. A. 1987b, Sov. Astr. AJ, 31, 517.

Weiler, K. W. and Sramek, R. A. 1988, Ann. Rev. Astr. Ap., 26, 295.

Weiler, K. W., Panagia, N., and Sramek, R. A. 1990, preprint.

Weiler, K. W., Sramek, R. A., Panagia, N., van der Hulst, J. M., and Salvati, M. 1986, Ap. J., 301, 790.

Low Frequency Spectral Lines from the Cold ISM

H. E. Payne
Space Telescope Science Institute
3700 San Martin Drive
Baltimore, MD 21218

The purpose of this contribution is simply to remind you that there are spectral lines at low radio frequencies. In contrast to all of the other astrophysics addressed by observations at low frequencies, these spectral lines are probes of cold, dense, and neutral regions in the interstellar medium.

The lines of interest are radio recombination lines. These lines have been observed at centimeter and millimeter wavelengths for twenty-five years, as discussed at a recent retrospective (IAU Colloquium No. 125) held in Pushchino, USSR. Decametric observations of recombination lines were first reported by Konovalenko and Sodin (1980), although the correct identification was made subsequently by Blake, Crutcher, and Watson (1980). The lines were seen in absorption, whereas all higher frequency lines had been seen in emission. Subsequent observations at intermediate frequencies have traced the transition between absorption and emission in at least one line of sight (Payne, Anantharamaiah, and Erickson 1989, hereafter PAE).

Recombination lines at decametric wavelengths arise from transitions in atoms where one electron is in a bound state with principle quantum number above 600. Such "hydrogenic" atoms have a classical radius of $20\,\mu\text{m}$ or larger, and are very sensitive to their surroundings. For this reason, the lines are a potentially important probe of the physical conditions in the regions where the lines originate. In this note I will review some of the properties of radio recombination lines, review some of the results that have been obtained from the ground, and describe the need for observations with a large array, perhaps in space.

Properties of radio recombination lines

In the interstellar medium, after a proton and an electron have recombined to form a hydrogen atom, that atom will usually have time for the electron to cascade down to the ground state. The most probable cascade involves changing the principal quantum number n by one at each step. For large n, the recombination line due to a transition from level $n + 1$ to n occurs at a frequency $\nu \sim n^{-3}$. There are many lines at low frequencies, since the number of lines per frequency interval increases faster than ν^{-1}. There are dozens of lines in the protected band centered near 24 MHz. If one line is being obliterated by interference, it is easy to tune to the next one, and a number of lines can be observed simultaneously if the spectrometer allows it. Lines from transitions with $n = 2$ are weaker by about a factor of seven, but have been observed in the decametric range.

Other atomic species having one electron in a highly excited state are called "hydrogenic," because the inner electrons shield all but a single unit of nuclear charge. The only difference between hydrogenic species is slightly different reduced masses, which introduces small frequency shifts. In practice, lines of hydrogen and helium are well separated, but everything heavier tends to blend together.

For levels n corresponding to centimeter wavelength recombination lines, the level populations are determined predominantly by radiative transitions. Radiative transition rates increase as n decreases, so the downward transitions out of a given level go faster than the transitions from higher levels that re-populate it. This leads to population inversions and stimulated emission. At some higher n, radiative transition rates will be so slow that level populations are determined predominantly by collisions with free electrons. In this case the level populations will be thermalized, and lines can be seen in absorption against a background source. The value of n at which the crossover between absorption and emission occurs depends upon the atomic physics, the electron temperature and density (which describe collisions), and the galactic non-thermal radiation field (whose photons induce radiative transitions).

The energy levels of a hydrogenic atom are perturbed by the Stark effect induced by passing electrons. This broadens the lines observed from an ensemble of atoms. In addition, the abundant photons of the non-thermal radiation field enhance radiative transition rates, and shorten the lifetime of an atom in a given state, which also broadens the observed lines. Both of these effects depend on n to a high power. To interpret the lines it is important to observe both broadened and unbroadened lines—the desire to observe broadened lines is a primary reason for observing at low frequencies. For a reasonable choice of parameters describing diffuse interstellar clouds, the discrete electronic states are expected to blend into the continuum at around $n = 1000$, which corresponds to about 6 MHz. There will be no lines in the spectrum at lower frequencies. Lines from regions with higher temperatures and higher electron densities, like HII regions or the warm ionized medium, will have been broadened away at much higher frequencies.

The interpretation of radio recombination lines depends on observations at a number of frequencies. The usual approach is to calculate line intensities, varying the electron temperature and density to match the observed variation of line intensity with frequency, and varying the thickness of the source region to get the normalization. Some simple geometry is assumed. One complication that cannot be ignored is the departure from thermodynamic equilibrium of the atomic level populations. For a given electron temperature and density, and galactic radiation field, a system of equations describing the balance between the populating and de-populating mechanisms for each level can be solved. Calculations relevant to interstellar conditions have been performed by Salem and Brocklehurst (1979), and Ungerechts and Walmsley (1979).

The recombination lines observed at low frequencies have been attributed to carbon atoms. This identification cannot be secured by the observed frequency shift, but is based, instead on the abundance of carbon and on having an ionization potential lower than hydrogen—carbon can be ionized (and recombining) in regions where the hydrogen is primarily neutral. Carbon is somewhat different from other hydrogenic atoms because there is fine structure splitting of the ground state in singly ionized carbon. The energy difference between the fine structure states is only 92 K; collisional excitation of this state, and the subsequent loss of a 156 μm photon is the dominant cooling mechanism in the neutral, atomic interstellar medium (Dalgarno and McCray 1972). Watson, Western, and Christensen (1980) first described a process similar to dielectronic recombination in which the fine structure state is excited while a low energy electron is captured into a bound state with large n. The capture is only temporary because the emission of a photon to stabilize the assemblage is too slow, but the effect on the level populations at large n and the resulting line intensities can be dramatic (Walmsley and Watson 1982a, 1982b).

Observational results

The first detection of spectral lines at low frequencies was made near 26 MHz (Konovalenko and Sodin 1980). Lines were observed in absorption against Cassiopeia A at frequencies corresponding to recombination lines from one of the heavier elements at the velocity of the Perseus arm. Blake, Crutcher, and Watson (1980) attributed the lines to carbon because the lines must originate in a region where the hydrogen is cold and primarily neutral, and carbon is the most abundant element with an ionization potential less than that for carbon. The fact that the lines have not been broadened away immediately implies low electron densities and temperatures.

Subsequent observations (*e.g.*, Konovalenko 1984, Ershov *et al.* 1984, Anantharamaiah, Erickson, and Radhakrishnan 1985, Ershov *et al.* 1987, Payne, Anantharamaiah, and Erickson 1989) have extended the frequency range of the Cas A detections and have detected other lines of sight. The hydrogen lines are not observed, so the hydrogen ionization rate and fractional ionization must be low. Most of the free electrons must come from carbon.

The Cas A line of sight remains the cardinal direction for a critical observational reason: observations at different frequencies are made with different beam sizes. The effect of beam dilution on a discrete source is complicated by the presence of the galactic non-thermal background. Only Cas A has an intensity which completely dominates that of the background, allowing observations with different beam sizes to be compared freely.

Payne, Anantharamaiah, and Erickson (1989), obtained spectra towards Cas A at frequencies between 325 MHz and 34 MHz, and observed the crossover from absorption into emission. The line was shown to consist of two distinct velocity features. There appears to be no optimum frequency for observing the lines in absorption; the lines are stronger at lower frequencies, but they must be observed with narrower bandwidths.

The current record for highest n is the spectrum obtained by Konovalenko (1990), which is the average of the C764α–C768α. Although the observed line width (nearly 200 km s^{-1}) is in agreement with observations at higher frequency, the broadening is so great that the

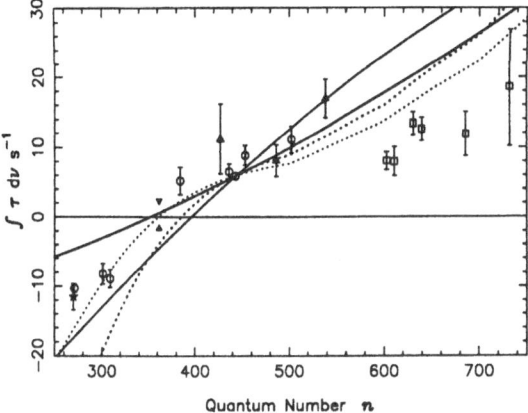

Figure 1: Integrated optical depth in the carbon recombination lines towards Cas A, as a function of principal quantum number n. The solid lines are predictions of the warm model and dashed lines are for cold models.

distinction between the two velocity features is completely obliterated. In the absence of any spatial resolution, observations at this low frequency (14.7 MHz) would add little to our understanding of this source if it were not for the higher frequency observations.

The current state of efforts to model the recombination line regions towards Cas A is illustrated in Figure 1, taken from PAE. The integrated optical depth under the two features is plotted as a function of principal quantum number. The curves are model calculations, which fall into two categories: *low temperature models*, colder than 20 K so that the dielectronic process is not active, and *warm models*, closer to 50–100 K so that the dielectronic process operates effectively. The hydrogen will be primarily neutral in the warm models but is presumably molecular in the low temperature models. The data do not allow us to distinguish between these quite different models.

Interferometric observations

In the hope of distinguishing between the two classes of models of the Cas A clouds, Erickson, Anantharmaiah, and Payne (1990) (see also Payne, Anantharamaiah, and Erickson 1990) observed Cas A at 332 MHz with the VLA at an angular resolution of about 2 arcmin, and good velocity resolution. These maps were compared to λ21 cm neutral hydrogen absorption maps (Greisen 1973, Schwarz *et al.* 1986), and maps of various molecular tracers, such as OH (Bieging and Crutcher 1986), H_2CO (de Jager *et al.* 1978), NH_3 (Batrla, Walmsley, and Wilson 1984), and CO (Troland, Crutcher, and Heiles 1985) to see if there were a stronger resemblance to neutral or molecular distributions.

The stronger of the two recombination line features corresponds to a heavily saturated feature in the neutral hydrogen maps. Still, there appears to be a general correspondence between maps in the velocity wings of this feature, while there is almost no molecular

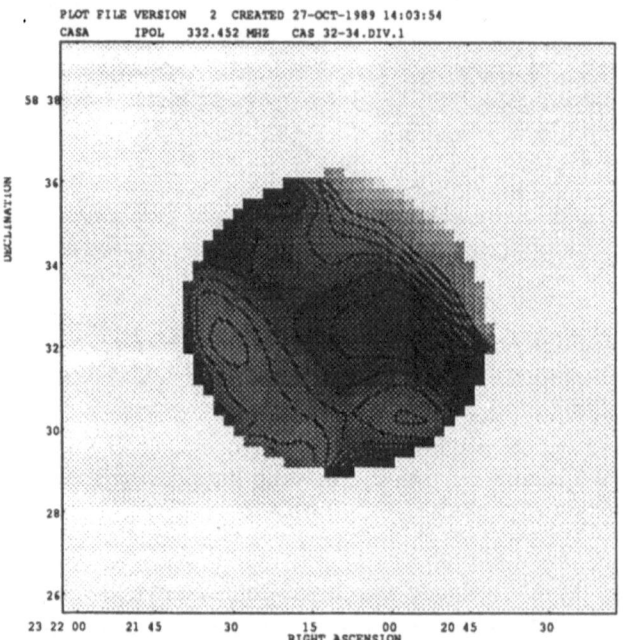

PLOT FILE VERSION   2   CREATED 27-OCT-1989 14:03:54
CASA   IPOL   332.452 MHZ   CAS 32-34.DIV.1

Figure 2: VLA map of carbon recombination line optical depth across Cas A, shown both in grey scale and contour map representations. Darker corresponds to higher optical depths. This is the average of three maps with velocities between -39.8 and $-37.0\,\mathrm{km\,s^{-1}}$.

material in one of these wings. The average of three optical depth maps spanning the weaker of the two recombination line features is shown in Figure 2 as a contour and grey scale plot. The feature seen extending from the center of the source to its western edge is seen in all tracers, but the feature seen in the northeastern edge of the source is seen only in the neutral hydrogen maps. Erickson, Anantharamaiah, and Payne (1990) argue that many of the recombination line features are seen in all tracers, but some are apparent only in the neutral hydrogen maps, indicating that the lines originate in neutral regions and that the warm models may be the most relevant. If the recombination lines and neutral hydrogen absorption lines originate in the same gas, then combining information from both observations provides a wealth of information, such as space densities, cloud sizes, carbon depletion, and heating and cooling rates in the gas, as described in PAE.

In this case the value of interferometric observations is clear. In fact, much higher resolution would be useful for comparison with the molecular maps. In addition, high resolution maps at lower frequencies would allow us to model individual features in the maps. The model shortcomings described in the previous section may be due to treating an ensemble of clouds with different physical conditions as a single, uniform cloud. Observations with an angular resolution of 0.5 arcmin at 75 MHz (VLA in A array) and at 25 MHz ($\sim 100\,\mathrm{km}$ baselines), in addition to better data at 332 MHz, would allow such detailed modelling.

In general, the value of interferometric observations would be to increase the number of discrete sources observed with comparable angular resolution at different frequencies. Modelling would then allow us to determine the physical conditions in a large sample of clouds, providing some very important information about this phase of the interstellar medium. No other method is likely to provide as direct a probe of the thermal balance in distant clouds as the combination of low frequency recombination lines and neutral hydrogen absorption lines.

Is there a need for observations from space? Observing lines down to the natural limit around $n = 1000$ would help to probe the interstellar radiation field, and to refine our understanding of the atomic physics of the carbon atom, but may add little to the understanding of the physical conditions in interstellar clouds. Observations at frequencies closer to 25 MHz would be more informative there, be they from space or from the ground.

REFERENCES

Anantharamaiah, K. R., Erickson, W. C., and Radhakrishnan, V. 1985, *Nature*, **315**, 647.

Batrla, W., Walmsley, C. M., and Wilson, T. L. 1984, *Astron. Astrophys.*, **136**, 127.

Bieging, J. H., and Crutcher, R. M. 1986, *Astrophys. J.*, **310**, 853.

Blake, D. H., Crutcher, R. M., and Watson, W. D. 1980, *Nature*, 287, 707.

Dalgarno, A., and McCray, R. A. 1972, in *Ann. Rev. Astron. Astrophys.*, **10**, 375.

de Jager, G., Graham, D. A., Wielebinski, R., Booth, R. S., and Gruber, G. M. 1978, *Astron. Astrophys.*, **64**, 17.

Erickson, W. C., Anantharamaiah, K. R., and Payne, H. E. 1990, in preparation.

Ershov, A. A., Ilyasov, Yu. P., Lekht, E. E., Smirnov, G. T., Solodkov, V. T., and Sorochenko, R. L. 1984, *Soviet Astron. Letters*, **10**, 348.

Ershov, A. A., Lekht, E. E., Smirnov, G. T., and Sorochenko, R. L. 1987, *Soviet Astron. Letters*, **13**, 8.

Greisen, E. W. 1973, *Astrophys. J.*, **184**, 363.

Konovalenko, A. A. 1984, *Soviet Astron. Letters*, **10** 353.

Konovalenko, A. A., and Sodin, L. G. 1980, *Nature*, **283**, 360.

Konovalenko, A. A. 1990, in *IAU Colloquim No. 125, Radio Recombination Lines*, ed. M. A. Gordon (Dordrecht: Reidel).

Payne, H. E., Anantharamaiah, K. R., and Erickson, W. C. 1989, *Astrophys. J.*, **341**, , 890 (PAE).

Payne, H. E., Anantharamaiah, K. R., and Erickson, W. C. 1990, in *IAU Colloquim No. 125, Radio Recombination Lines*, ed. M. A. Gordon (Dordrecht: Reidel).

Salem, M., and Brocklehurst, M. 1979, *Astrophys. J. Suppl.*, **39**, 633.

Schwarz, U. J., Troland, T. H., Albinson, J. S., Bregman, J. D., Goss, W. M., and Heiles, C. 1986, *Astrophys. J.*, **301**, 320.

Troland, T. H., Crutcher, R. M., and Heiles, C. 1985, *Astrophys. J.*, **298**, 808.

Ungerechts, H., and Walmsley, C. M. 1979, *Astron. Astrophys.*, **80**, 325.

Walmsley, C. M., and Watson, W. D. 1982a, *Astrophys. J.*, **255**, L123.

Walmsley, C. M., and Watson, W. D. 1982b, *Astrophys. J.*, **260**, 317.

Watson, W. D., Western, L. R., and Christensen, R. B. 1980, *Astrophys. J.*, **240**, 956.

HII REGIONS IN ABSORPTION AT LOW FREQUENCIES

Namir E. Kassim
ONT Postdoctoral Fellow
Center for Advanced Space Sensing
Naval Research Laboratory
Washington, DC 20375-5000

ABSTRACT

This paper reviews the information provided by the analysis of low radio frequency ($\nu < 100$ MHz) observations of HII regions seen in absorption against the Galactic background radiation. We show how such observations can be used to constrain various physical parameters associated with the Galactic background and the HII regions. The technique is illustrated by analysis of 30.9 MHz observations of the W51 giant HII complex. The importance of obtaining observations made at more than one low frequency is emphasized.

INTRODUCTION

The absorbing property of ionized gas in the interstellar medium (ISM) at low radio frequencies is a powerful tool for Galactic astrophysics. Dulk and Slee (1972, 1975) and Kassim (1989) have used the low frequency turnovers in the spectra of Galactic supernova remnants (SNRs) to constrain the distribution of ionized gas in the ISM responsible for the absorption. Kassim and Weiler (1990) have used the appearance of an HII region in absorption against the SNR G8.7-0.1 to set a lower limit to its distance and thus suggest a possible association with the very young pulsar PSR1800-21. In this paper, we explore the information obtained from observing HII regions in absorption against the Galactic background that is applicable to observations in the 10-100 MHz regime. We review an analysis of 30.9 MHz observations of W51 by Kassim (1987,1988) as an example of the application of the technique. These results are then compared to independent estimates of the derived parameters made by other observers. Reynolds (see paper in these proceedings) has suggested that the techniques we describe below may also be used to explore the properties of the warm ionized medium (WIM) component of the ISM at even lower radio frequencies (<5 MHz).

OBSERVATIONS

Figs. 1 and 2 show 30.9 MHz continuum maps from the Clark Lake Galactic plane survey (Kassim 1988). Notice the large number of deep absorption regions (black holes) located towards the inner Galaxy, and note how their effects decrease with increasing Galactic longitude. Rarely would such absorption regions be seen in areas other than towards either the first or fourth Galactic quadrants.

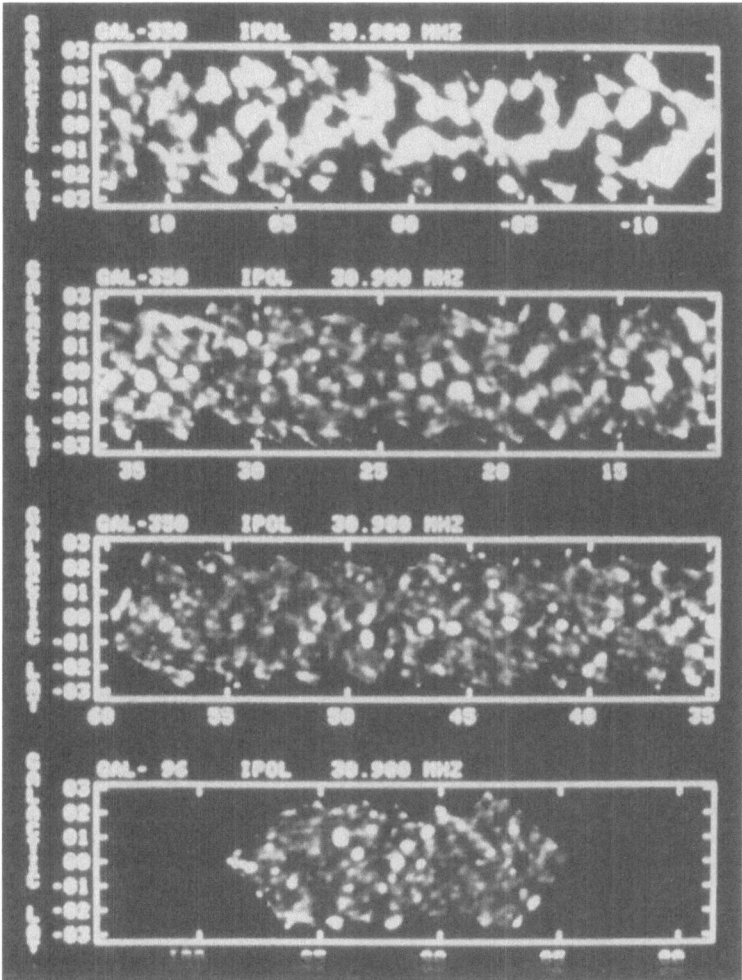

Fig. 1: Grey scale photograph of 30.9 MHz continuum maps of the inner Galaxy taken from the Clark Lake Galactic Plane Survey (Kassim 1988). Each strip is centered on the Galactic equator and extends over Galactic latitude ~b = ±2.5°. Galactic longitude increases from right to left and each strip covers approximately as follows: Strip #1: l=-12° to +12°; Strip #2: l=10° to 36°; Strip #3: l=35° to 60°; Strip #4: l=85° to 98°. The resolution of the "T" shaped array used to make the maps is approximately 15 arcminutes at the zenith. Dark "holes" concentrated towards the lowest longitudes are discrete absorption holes produced by foreground HII regions. The W51 HII region complex appears as a depression near l=49°, b=-0.5° on Strip #3.

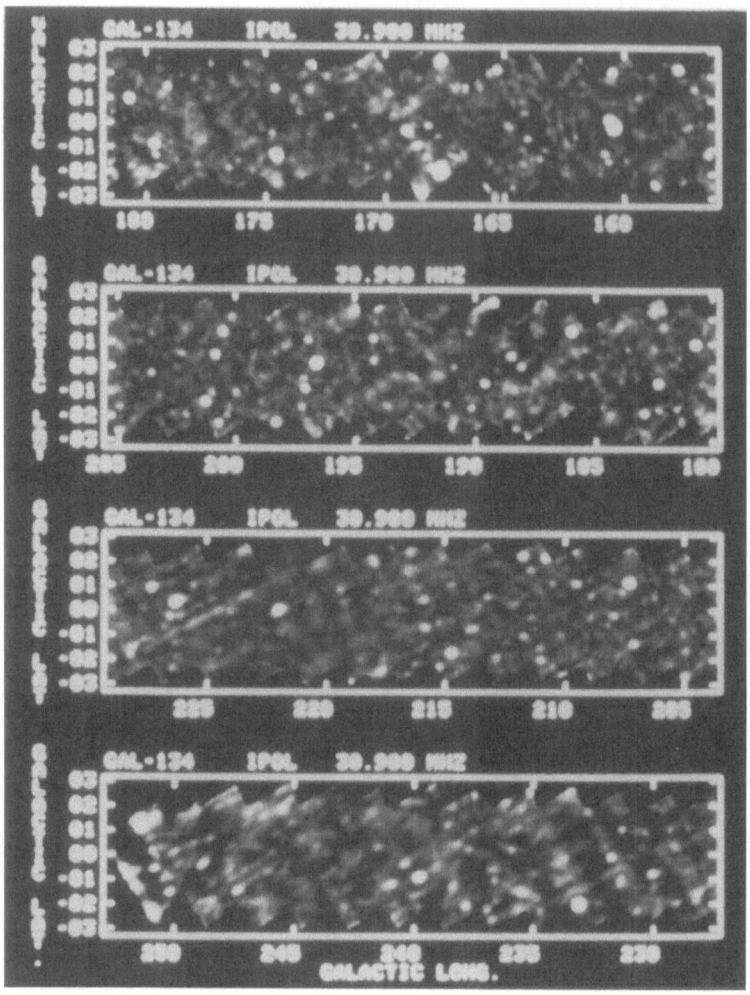

Fig. 2: Grey scale photograph of 30.9 MHz continuum maps made towards the outer Galaxy taken from the Clark Lake Galactic Plane Survey (Kassim 1988). Strip scans are the same as in Fig. 1 except that the coverage in Galactic longitude is as follows: Strip #1: l=155° to +181°; Strip #2: l=180° to 205°; Strip #3: l=204° to 229°; Strip #4: l=227° to 252°. Source counts (Kassim 1987) confirm that most of the sources seen here are extragalactic. (The small region beyond l=250° is distorted by edge effects and should be ignored.)

SIMPLE ANALYSIS

Fig. 3 is a diagram that illustrates the relevant geometry. The telescope is pointed towards the inner Galaxy, and a foreground (i.e., nearby compared to the distance to the Galactic center) HII region is located along the line-of-sight. In a simple analysis, we can characterize the HII region as having an electron

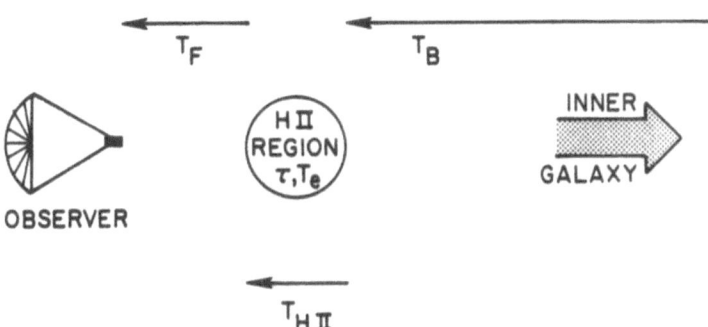

Fig. 3: Simplified HII Region Absorption Geometry

temperature T_e and free-free optical depth τ. We define its brightness temperature to be T_{HII}. We divide the emission due to the extended Galactic background radiation into two components. These are T_B and T_F which are, respectively, the components of the synchrotron emissivity originating from beyond (<u>B</u>ackground) and from in front (<u>F</u>oreground) of the HII region. Finally, we define T_T as the sum of T_B and T_F, i.e., the total brightness due to the distributed Galactic emission.

Now consider the simple case of a large, optically thick HII region with optical depth $\tau \gg 1$, i.e. where $T_{HII} = T_e$. If this HII region is observed by a single-dish telescope with infinite resolution, then Fig. 4 illustrates the picture that emerges as one scans the telescope beam across the Galactic plane.

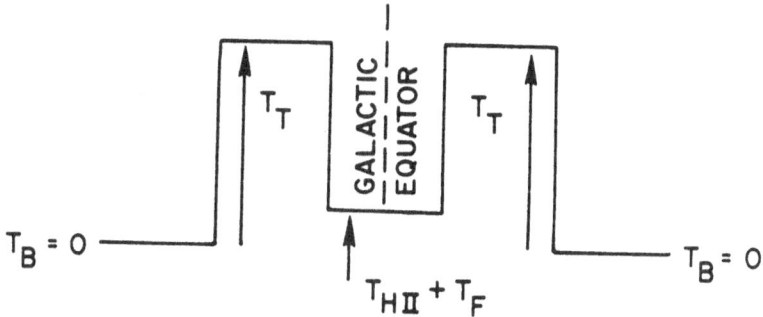

Fig. 4: Simplified illustration of single-dish scan across HII region.

Now, at 30.9 MHz we have roughly that $T_T \sim 100,000$ K, $T_{HII} = T_e \sim 10,000$ K, and $T_F \sim 10,000$K. Therefore when the telescope beam is on the HII region, the observed brightness temperature is $T_{HII} + T_F \sim 20,000$K. But since the observed brightness temperature when the beam is no longer on the HII region is T_T, which is often much larger than 20,000K, the HII region appears as a <u>depression</u> on the map.

If the same observation were conducted with an interferometer which has no zero-spacing, i.e., it cannot measure T_T, then the scan would like Fig. 5. In this case the HII region actually appears as

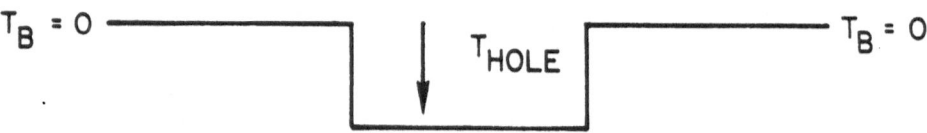

Fig. 5: Simplified illustration of interferometer scan across HII region.

a "HOLE" on the map. If we define the observed brightness temperature when the telescope beam is on the HII region to be T_{HOLE}, then we have:

$$T_{HOLE} = T_{HII} + T_F - T_T = T_{HII} + T_F - [T_F + T_B] \tag{1}$$

and therefore

$$T_{HOLE} = T_{HII} - T_B < 0 \tag{2}$$

THEORY

In general, the brightness temperature of the depression T_D observed towards an optically thick HII region by an interferometer at some position in the sky (α, δ) at some low frequency ν is given by:

$$T_D(\alpha, \delta, \nu) = [T_F + T_B e^{-\tau} + T_e(1 - e^{-\tau})] - T_T \tag{3a}$$

$$= (T_e - T_B)(1 - e^{-\tau}) \tag{3b}$$

or in the most general case where T_x is the brightness contribution from the extragalactic background and $P(\alpha, \delta)$ is the normalized polar diagram of the antenna power pattern:

$$T_D(\alpha, \delta, \nu) = [T_e - T_B(\nu) - T_x(\nu)] \times \int \{1 - \exp[-\tau(\alpha + \alpha', \delta + \delta', \nu)]\} \times P(\alpha', \delta') d\alpha' d\delta' \tag{4}$$

In the simplest case, if we ignore the antenna pattern of the telescope (i.e., assume infinite resolution) and assume $\tau \gg 1$, we arrive at the following basic (simplified) relationship:

$$T_D = [T_e - T_B - T_x] \tag{5}$$

Next we consider the practical application of these equations to low frequency observations.

METHOD OF ANALYSIS

First we consider the analysis of data based upon observations made at only one low frequency. In the simplest case where $\tau \gg 1$, we can make a rough estimate of the electron temperature T_e of the HII region which is often known from measurements made at higher frequencies. If we then measure a depression of brightness temperature T_D, we can solve for T_B from equation (3b). If we also have available a measurement of T_T (i.e. if we can measure a zero spacing flux), we can solve for both T_B and T_F, since $T_T = T_F + T_B$. If in addition, we know the distance D to the HII region, we can estimate the the Galactic emissivity E_ν originating along the line-of-sight to the HII region, i.e., E_ν (K/pc)=T_F(K)/D(pc). In the next simplest case, we could again assume T_e but estimate τ from either: (1) higher frequency recombination line observations; or (2) from optically thin continuum maps where the approximation $T_{brightness} \sim \tau T_e$ is valid. Both methods make use of reliable measurements which are currently readily available in the literature.

Now let us further consider analysis of measurements of T_D made at two sufficiently spaced low frequencies (say 30 and 60 MHz). If we measure T_D and T_T, and we know the distance to the source, we can solve for T_B, T_F, the foreground Galactic emissivity E_ν and T_e. That is, we need only get τ from the higher frequency data. This result can be generalized as follows: HII region absorption "HOLES" observed at low frequencies depend on a number of interesting parameters. With only a single measurement, most of these must either be assumed or estimated from higher frequency data and, if appropriate, be extrapolated to the lower frequency. If one is able to make more than one low frequency measurement, the dependence on either assumptions or results obtained from higher frequency data is reduced.

APPLICATION TO 30.9 MHZ OBSERVATIONS OF W51

The HII region W51 appears as a depression on Fig. 1 near l=49°, b=-0.5°. The following simplified analysis summarizes results originally presented by Kassim (1987). From analysis of Fig. 1, T_{HOLE} = - 35,000 K at 30.9 MHz towards W51. Also, the assumption of $\tau \gg 1$ is supported by both recombination line data and optically thin continuum measurements (Wink et al. 1983; Downes et al. 1984, Altenhoff et al. 1978). Therefore from equation (5) we have $T_{HOLE} = T_e - T_B - T_x = -35,500$ K. Recombination line measurements (Wink et al. 1983; Downes et al. 1980) give $T_e \sim 7,500$ K, and we can use the model by Bridle (1967) to estimate $T_x \sim 3,700$ K. Therefore, equation (5) gives $T_B = 39,300$ K $\pm 4,000$ K. This result compares well with results of Desphande and Sastry (1986) based on 34.5 MHz observations using the Gaiibidinauer telescope of T_B (scaled to 30.9 MHz) = 37,100 K $\pm 3,000$ K.

Next, we can estimate T_T from "old" low frequency surveys. Using an average of values estimated by Milogradov-Turin and Smith (1974), Parish (1972) and Cane (1978) we find $T_T \sim 54,100$ K. This leads directly to $T_F = 14,800$ K (from $T_T = T_F + T_B$). Finally if W51 is at 6.5 kpc (Beiging 1975), we can solve for the line-of-sight synchrotron emissivity E_ν=2.3 K/pc. This result again compares well with results of Deshpande and Sastry (1986) (2K/pc) and of Milogradov-Turin (1974) (2-3 K/pc).

SUMMARY

At low radio frequencies, foreground HII regions seen in absorption towards the inner Galaxy will make the Galactic Plane look like "Swiss cheese". Measurements of these holes can be combined with a variety of higher frequency data to constrain the following physical parameters: electron temperature of the HII Region (T_e), free-free optical depth of the HII Region (τ), brightness temperature of the distributed Galactic synchrotron emission originating from behind (T_B) and from in front (T_F) of the HII region, and the synchrotron emissivity (E_ν) originating along the line-of-sight to the HII region. When observations are available at more than one low frequency, the dependence on either assumptions or extrapolations based on higher frequency observations are reduced.

E_ν is particularly important to cosmic ray physics because cosmic rays are intimately related to the distributed Galactic synchrotron emission via their interaction with the Galactic magnetic field. In particular, the synchrotron emission in the 1-30 MHz frequency range is related to cosmic rays in the 100 MeV - 1 GeV energy range (respectively) which cannot be directly measured near the Earth due to the modulation of the solar wind. (This topic is covered in more detail in the articles by Longair and Webber in these proceedings.)

In addition we note that the techniques described in this paper can also be used to resolve the HII region distance ambiguities (determined kinematically) because only "near" (i.e. foreground) HII regions would appear in absorption against the Galactic background at low frequencies.

Finally we point out that Reynolds (see paper in these proceedings) has suggested that the principles employed above can be extended to even lower radio frequencies in order to explore the absorption properties of the important but very poorly understood warm ionized medium (WIM) component of the ISM.

REFERENCES

Altenhoff, W.J., Downes, D., Pauls, T., and Schraml, J. 1978, Astr. Ap. Suppl., **35**, 23.

Bieging, J. 1975, in "HII Regions and Related Topics", eds. T.L. Wilson and D. Downes, (Berlin:Springer), pg. 443.

Bridle, A. H. 1967, M.N.R.A.S., **136**, 219.

Cane, H.V. 1978, Aust. J. Phys., **31**, 561.

Deshpande, A.A. and Sastry, C.V. 1986, Astr. Astrophys., **160**, 129.

Downes, D., Wilson, T.L., Beiging, J., and Wink, J. 1980, Astr. Ap. Suppl., **40**, 379.

Dulk, G.A. and Slee, O.B. 1972, Australian J. Phys., **25**, 429.

Dulk, G.A. and Slee, O.B. 1975, Ap. J., **199**, 61.

Kassim, N.E. 1987, Ph.D. thesis, University of Maryland.

Kassim, N.E. 1988, Ap. J. Suppl., **68**, 715.

Kassim, N.E. 1989, Ap. J., **347**, 915.

Kassim. N.E. and Weiler, K.W. 1990, Nature, **343**, 146.

Longair, M.S. 1990, in these proceedings

Milogradov-Turin, J. 1974, Mem. Soc. Astron. Ital., **45**, 85.

Milogradov-Turin, J. and Smith, F.G. 1974, M.N.R.A.S., **161**, 269.

Parrish, A. 1972, Astron. J., **174**, 33.

Reynolds, R.J. 1990, in these proceedings.

Webber, W.R. 1990, in these proceedings.

Wink, J.E., Wilson, T.L., and Beiging, J.H. 1983, Astr. Ap., **127**, 211.

V. SCATTERING IN THE ISM AND PULSAR OBSERVATIONS

LOW-FREQUENCY ANGULAR BROADENING AND DIFFUSE INTERSTELLAR PLASMA TURBULENCE

Steven R. Spangler
Department of Physics and Astronomy, University of Iowa
Iowa City, Iowa 52242, U.S.A.

John W. Armstrong
Jet Propulsion Laboratory, 4800 Oak Grove Drive
Pasadena, California 91109, U.S.A.

Abstract

A low-frequency radio interferometer is potentially a powerful tool for studies of diffuse plasma turbulence. The scattering angular size of a source due to interstellar scattering is proportional to the square of the observing wavelength. Synchrotron self-absorption, an intrinsic mechanism for making a source larger at lower frequencies, produces an angular size proportional to the first power of the wavelength. Thus, observations at a longer wavelength will always be more sensitive to scattering than observations at higher frequencies. In this article we consider in detail two issues. (1) The effect of turbulence in the *interplanetary* medium is considered. We conclude that interplanetary scattering will limit the effective sensitivity of a low-frequency interferometer to far above its theoretical value. Any serious design considerations for a low-frequency, space-borne interferometer must include a detailed investigation of the effect of the interplanetary medium. (2) If the limitations imposed by the interplanetary medium can be overcome, a low-frequency interferometer could be used to search for turbulence near supernova remnants. Such turbulence has been hypothesized to play an important role in the acceleration of the cosmic rays, and observational evidence for its existence would be a major accomplishment. Finally, we discuss a couple of additional, more minor topics which could be addressed with such an instrument, such as studies to verify if the low-frequency variability of extragalactic radio sources is a scintillation phenomenon.

1 Introduction and Definition of Scattering Parameters

An electromagnetic wave propagating through a turbulent plasma ("turbulent" here meaning that the plasma has a density which varies spatially in a random manner) is scattered into angles different from the initial angle of propagation. A source viewed through such a medium will appear "blurred" or broadened to an angular size considerably larger than its intrinsic size. The observed angular size, θ_0, will approximately be given by,

$$\theta_0 = \sqrt{\theta_I^2 + \theta_s^2} \, , \tag{1}$$

where θ_I is the intrinsic size, i.e., that which would be observed in the absence of the medium, and θ_s is the "blurring size" imposed by the medium. Each of these angular sizes depends, in general, on the wavelength of observation.

Most of the work we will be considering involves observation of extragalactic radio sources, in which we use these objects as probes of the interstellar medium. For our purposes then, θ_I will refer to the angular size of a synchrotron radiation source. If the source is observed at a frequenncy at which it is entirely optically thin, typically several GHz or higher, θ_I should be independent of wavelength. At frequencies at which the source is partially opaque, usually frequencies below a few GHz, $\theta_I \propto \lambda$ (Marscher 1977). The scattering angular size $\theta_s \propto \lambda^x$, where $x \simeq 2$, with the precise value of x depending on the form of the density power spectrum. These elementary considerations indicate that the lower the frequency of observation, the larger the contribution of the scattering angular size to the observed size. The current best tool available to astronomers is intercontinental VLBI at 610 or 327 MHz. Such observations are severely affected by the ionosphere, thus supporting the arguments for an interferometer in space.

We now consider the relation between the scattering size, θ_s, and the properties of the turbulence. The density irregularities in a turbulent medium may be characterized by their spatial power spectrum. In the case of the interstellar medium, this spatial power spectrum is taken to be a power law,

$$P_{\delta n}(k) = C_N^2 k^{-\alpha}, \quad k_0 < k < k_i \, , \tag{2}$$

where k_0 and k_i are the reciprocals of the outer and inner scales, respectively. The spectrum (2) is therefore characterized by two parameters, the power law exponent α and the normalization constant C_N^2.

A considerable amount of recent observational work indicates that the value of $\alpha \cong 3.67$, the "Kolmogorov" value (see Cordes, Rickett, and Backer 1988). It is also clear that C_N^2, a measure of the "intensity" of the turbulence, varies drastically from point to point in the interstellar medium although apparently α does not. For a given value of α, the strength of scattering is determined by $C_N^2 Z$, where Z is the distance the wave has propagated through the turbulent medium. For a Kolmogorov spectrum and a fiducial observation frequency of 1 GHz, the value of θ_s (FWHM) is (Spangler et al. 1986)

$$\theta_{1\,\text{GHz}} = 2.24(C_N^2 Z_{\text{pc}})^{0.6} \quad \text{mas} \, . \tag{3}$$

In equation (3) the path length through the medium is given in parsecs, while C_N^2 is in units of $\text{m}^{-20/3}$.

Equation (3) is represented in graphical form in Figure 1 as the solid diagonal line. The plotted points represent available, published broadening measurements of scattering or upper limits thereto, made with existing VLBI interferometers. The sources chosen form a representative, not exhaustive, sample. The source 2013+370 represents an easy detection of scattering with an interferometer operating at centimetric wavelengths with baselines confined to the United States (Spangler and Cordes 1988). The sources 2050+364 (Mutel and Hodges 1986) and 0016+731 (Spangler *et al.* 1986) represent, respectively, a detection and an upper limit with an intercontinental interferometer operating at a frequency of 610 GHz.

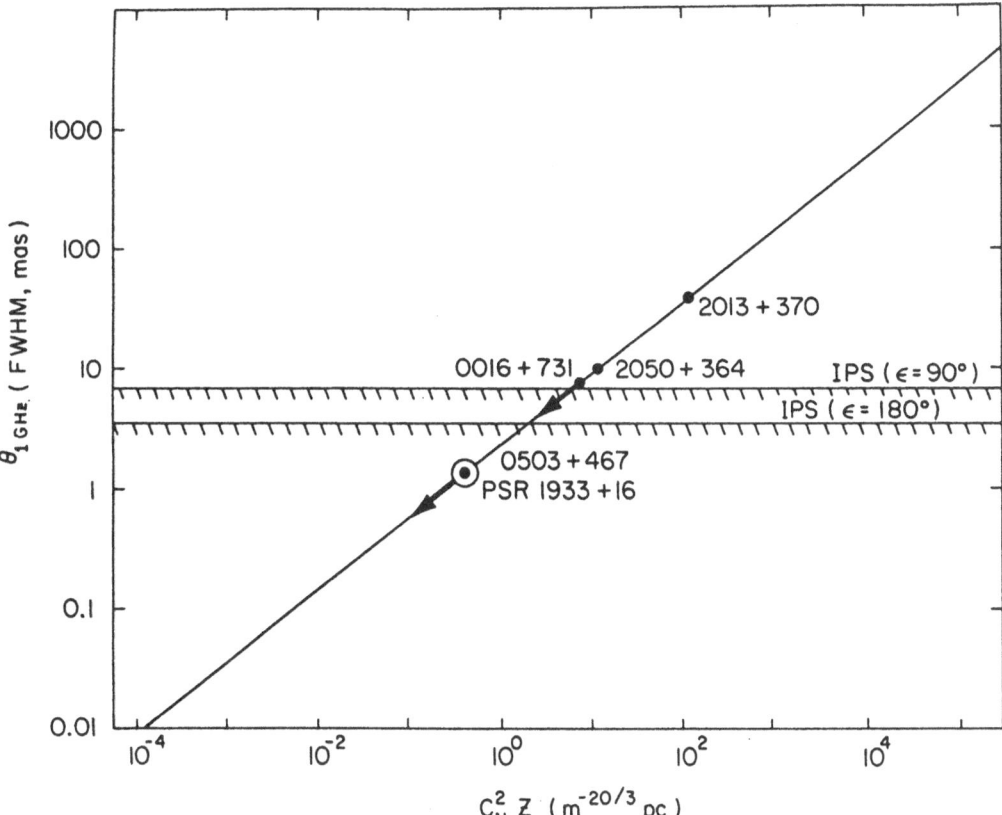

Figure 1. A plot of the relationship between the observable scattering angular size (at a fiducial frequency of 1 GHz) and the scattering measure $C_N^2 Z_{pc}$. The solid diagonal line represents the relation (3) expected for a Kolmogorov spectrum of density turbulence. The labeled data points represent existing published measurements made with VLBI interferometers. A space-borne interferometer operating at a frequency of 10 MHz would have the potential of resolving sources with an equivalent 1 GHz angular size of 0.01−0.1 mas, a significant improvement on existing VLBI interferometers. The horizontal hatched lines represent the effective angular broadening due to *interplanetary* density irregularities for solar elongation of 90° (upper line) and 180° (lower line).

The most extreme measurements in this assortment are an upper limit for the extragalactic radio source 0503+467 (Spangler, Fey, and Cordes 1987) and a measurement of scattering for the pulsar PSR 1933+16 (Gwinn, Moran, and Reid 1988). These data

data probably represent the limit for current and near-future VLBI interferometers. By the same token, many lines of sight to pulsars and extragalactic radio sources could be probed to the same sensitivity in $C_N^2 Z_{pc}$ as has been the case for 0503+467 and PSR 1933+16.

A low-frequency interferometer in space could, in principle, considerably extend our sensitivity in $C_N^2 Z_{pc}$. By way of example, let us consider an interferometer operating at a frequency of 10 MHz. We assume sensitivity and calibration of the interferometer are such that a drop in the visibility to 0.75 on the longest baselines can be confidently measured. If the longest baseline is 500 km, a source with a Gaussian-equivalent full-width at half-maximum size of 3''5 could be resolved. If the longest interferometer baseline were 5000 km, the minimum resolvable angular size would be 0''35.

The significance of these numbers for Figure 1 is as follows. Source resolution of the sort described above would be interpreted as angular broadening, or an upper limit thereto. These angular sizes may be transcribed to Figure 1 after scaling by the frequency ratio, 10 MHz to 1 GHz to the 2.2 power, which is appropriate for a Kolmogorov density spectrum. The corresponding scattering sizes at 1 GHz are 0.14 mas for a longest baseline of 500 km, and 0.014 mas for a 5000 km maximum baseline. We may see from Figure 1 that the ability to detect such small scattering angles would significantly extend the accessible range of scattering measure $C_N^2 Z_{pc}$.

These calculations demonstrate the desirability of a low-frequency space interferometer from the standpoint of interstellar scattering. Such an instrument could in principle detect much smaller scattering angles than present VLBI instruments. This observational capability would open the prospect of measuring extremely weak turbulence distributed over large path lengths, or strong turbulence restricted to small regions. Examples of the former class of turbulence might be hydromagnetic waves in interstellar tunnels, while the latter might include strong turbulence in the vicinity of shock waves. Either case would be of considerable astrophysical interest.

However, to realize this potential will require the envisioned interferometer to operate at close to optimum performance. In the next section we consider an effect which might severely impede the performance of the proposed interferometer.

2   Interplanetary Scattering—Can We See Out?

Sources observed for purposes of studying the interstellar medium are also viewed through the interplanetary medium, another turbulent plasma. For the purposes of the proposed low-frequency interferometer, the interplanetary medium plays the same role as does the Earth's atmosphere for optical telescopes; it limits the attainable resolution.

Indications that the interplanetary medium can substantially limit the resolution of low-frequency interferometry goes back to the work of Erickson (1964), who found that a variety of low-frequency angular broadening data could be fit with the following simple formula,

$$\theta \cong 50 \left(\frac{\lambda_m}{P_\odot}\right)^2 \quad \text{arcminutes} , \tag{4}$$

where θ is the scattered angular size due to turbulence in the solar wind, λ_m is the wavelength of observation (meters) and $P_\odot$ is the "impact parameter" between the line of sight and the Sun, in solar radii. Equation (4) was formulated on the basis of measurements out to solar elongations of about 25°. The Erickson formula would indicate a 10 MHz scattering angle of ~ 4.5 at an elongation of 25°. This is vastly greater than the potential resolving power of the interferometer discussed above, and as such would represent a major limitation of the performance of the instrument.

This calculation, of itself, only indicates that observations with the proposed interferometer could not be made close to the Sun. However, in view of the large amount of scattering given by equation (4), there arises the question as to whether there is *any* part of the sky in which the interplanetary scattering is sufficiently small as to permit effective utilization of the interferometer for scattering studies.

Since direct observations do not exist, we are forced to rely on extrapolations resulting from other types of scintillation observations. To quantitatively assess scattering in the interplanetary medium, we applied a model of the solar wind created to evaluate the effect of interplanetary scintillations on the Solar Probe telemetry link (Armstrong and Woo 1980). Almost all scattering observables depend on the power spectrum of the electron density irregularities. The Armstrong-Woo model assumes a spectrum of the form in equation (2), with the value of C_N^2 depending on the heliocentric distance. The model does not attempt to describe azimuthal, latitudinal, or temporal variations in C_N^2, but does correctly account for subtleties such as the variation of power law index with distance closest to the Sun (Woo and Armstrong 1979). Normalization of the model was based on scintillation index observations of compact radio sources and spacecraft beacons close to the Sun. To within a factor of about 15%, the model correctly reproduces observations at X, S, and L band taken in the elongation range $10-60\,R_\odot$ (Coles 1979, Armstrong and Woo 1981, Bourgois and Cheynet 1972).

For the case of interest here, in which the lines of sight to the sources always pass more than $20\,R_\odot$ from the Sun, the spectrum is Kolmogorov. The radial dependence of C_N^2 is given by

$$C_N^2 = 3.9 \times 10^{15}(R/R_\odot)^{-4.05} \quad \mathrm{m}^{-20/3} \; , \tag{5}$$

where R is the distance from the Sun, and $R_\odot$ is the solar radius (7×10^{10} cm).

Given the form (5), the scattering measure for an arbitrary line of sight can be calculated by numerical integration and used to calculate the interferometric visibility using equation (9) of Rickett (1977).

The results of such a calculation show that low-frequency observations with modest baseline lengths give low correlation, even in the antisolar direction. As a representative example, at an observing frequency of 10 MHz and a solar elongation of 180°, the visibility for an intrinsic point source will fall to e^{-1} on a baseline of only 40 km. The visibility is less than 1% for baselines greater than 100 km. This corresponds to an angular size (Gaussian FWHM) of 1.4 arcminutes. We can scale this angular size to our fiducial frequency of 1 GHz by having the scattering size be proportional to $\lambda^{2.2}$, the appropriate scaling for a Kolmogorov spectrum of irregularities. The corresponding $\theta_{1\,\mathrm{GHz}}$ is 3.3 mas. At a solar elongation of 90°, the antenna separation corresponding to a visibility of 0.37 at 10 MHz is only 24 km. The scattering size θ_1 GHz at $\epsilon = 90°$ is 5.5 mas. These limits are indicated by the horizontal hatched lines in Figure 1.

It will be noted that interplanetary scattering limits the resolution of a low frequency space interferometer to a level one to two orders of magnitude worse than the theoretical resolution. An immediate objection to Figure 1 and the model which produced it is that our predicted scattering size due to the interplanetary medium is larger than *observed* angular sizes for the sources 0503+467 and PSR 1933+16. The model predicts heavier scattering than has been observed for other objects as well.

By way of support for the credibility of this model for the solar wind turbulence, we cite the following three points of agreement with observation.

(1) Interplanetary intensity scintillations of the Crab pulsar (a point source for the purpose of this calculation) made at $1 = 4$ m, give intensity scintillation indices (rms intensity/mean intensity) of about 0.2 in the antisolar direction. The model above predicts 0.3, in excellent agreement.

(2) Interplanetary phase scintillation measurements (made using the coherent S- and X-band transmissions from deep space probes) have been made over a variety of elongation angles by Woo and Armstrong (1979). Spectra of spacecraft phase measurements observed at 175° elongation agree within about a factor of 2 with the spectra predicted from the above model.

(3) Crustal dynamics VLBI data taken on extragalactic sources at elongations of about 20 $R_\odot$ (observing at 0.12 m wavelength on $\sim$ 5000 km baselines) show phase variances of 50 radians^2 (Coles and Harmon 1988). Scaling these data appropriately in elongation, observing wavelength, and baseline length yields an estimate of the baseline for $1/e$ visibility of 70 km (30 m wavelength, 90° elongation). This compares with about 24 km calculated above, in fairly good agreement.

Our resolution of this apparent contradiction is as follows. The angular size calculated by our model and displayed in Figure 1 is an *ensemble-average* angular size, *i.e.*, that obtained by averaging the visibility over a period of time sufficient for the observational average to closely approximate the average obtained from a large number of independent measurements. This matter has recently been discussed by Goodman and Narayan (1989). For the low-frequency VLBI observations of 0503+467 and PSR 1933+16, short integration times and use of phase closure techniques removed the interferometer phase fluctuations which would produce loss of coherence. In essence, the effect of the interplanetary medium was removed. A striking demonstration of the removal of the effect of a scattering medium, given sufficiently high temporal and spectral resolution, has recently been presented by Cornwell, Anantharamaiah, and Narayan (1988).

The effect of the interplanetary medium on the proposed low-frequency array then becomes a matter of the obtainable temporal and spectral averaging. If the sensitivity of the instrument is sufficiently low, so that long integration times are required to detect most sources, the instrument will be operating in "ensemble-average" mode, and the horizontal lines in Figure 1 will be in effect. In this case, the low-frequency space interferometer will be *less* effective than current VLBI interferometers for scattering studies.

If, however, sufficient signal to noise exists to detect the IPS phase fluctuations, phase closure techniques may be employed and the interplanetary "seeing disk" substantially reduced.

The main conclusion of this section is that a thorough study of the effects of the interplanetary medium will be required before the design of a low-frequency interferometer progresses very far.

3   Interstellar Scattering as a Diagnostic of Shock-Associated Turbulence

One of the areas in which substantial progress has recently been made in space plasma physics is in studies of collisionless shock waves. Examples are planetary bow shocks, traveling interplanetary shock waves, and strong interaction regions between comets and the solar wind. A common set of phenomena characterize all of these shocks (at least the *quasi-parallel* shocks in which the upstream magnetic field lies perpendicular to the shock surface). These phenomena include reflected ions, which "bounce" from the shock, upstream hydromagnetic waves excited by these ions, and an ion population which has been isotropized by the self-excited waves and energized via repeated interactions with the shock front (Lee 1982). Observations of these phenomena have provided the experimental basis of the first- and second-order Fermi particle acceleration theories. The first-order Fermi acceleration mechanism is at the present the favored one for the origin of the galactic cosmic rays. It and the second-order process have been proposed as the acceleration mechanism for relativistic electrons in supernova remnants and extragalactic radio sources.

It would be a considerable accomplishment if we could detect shock-associated hydromagnetic waves in association with supernova remnants, which are known to be powerful interstellar examples of shock waves, and which in some cases are known accelerators of electrons. Such a method has been proposed by Spangler et al. (1986) and Spangler, Fey, and Cordes (1987), and utilizes the technique of interstellar scattering. Briefly, the hydromagnetic waves excited near a shock are compressive; observations near the Earth's bow shock indicate that the density modulation index, defined as the standard deviation of the density divided by the mean density, is 10–20%, and arguments could be summoned that a similar value would be appropriate for interstellar shocks. These density perturbations could produce radio wave scattering effects. The question naturally arises as to whether detectable scattering would result from these density fluctuations. This matter is discussed in Spangler et al. (1986) and Spangler, Fey, and Cordes (1987). Briefly, the upstream wave layer will be quite thin, $\sim 10^{-2}$ parsecs, unless it is inflated by cosmic ray generated waves, which is quite plausible. However, even for such thin upstream layers, a long glancing path at the edge of the remnant could pass through several tenths of a parsec of scattering material. This estimate is based on an expression for the path length through the region upstream of a supernova remnant presented in Spangler, Fey, and Cordes (1987). For such a condition, the product $C_N^2 Z_{\rm pc}$ would approach unity. From equation (3) we see that such a degree of scattering would barely be detectable with observations of extragalactic radio sources by current interferometers, and would be well within the capability of a space-borne, low-frequency interferometer. Successful observations of this sort would require a high density of observable extragalactic radio sources. The ability to make VLBI observations of sources interestingly close to supernova remnants has required Mark III VLBI, which permits observations of sources at flux levels of tens of milliJanskys. If the proposed array is only sensitive enough to observe a couple of hundred sources, it is unlikely that any will fortuitously lie close enough to a supernova remnant to allow these types of measurements.

4 Selected Additional Topics

The proposed low-frequency interferometer could also be used to address a number of additional topics, which we briefly discuss below.

Inner Scale for the Diffuse Interstellar Plasma. A precise measuerment of the interferometric visibility function of a source can be used to measure the inner scale in a turbulent medium, *i.e.*, the scale on which the density spectrum (2) is truncated. Briefly, if the spectrum is Kolmogorov or nearly so, and the separation of the antennas is larger than the inner scale, the visibility of a point source is given by

$$V(r) = \exp(-Ar^{\alpha-2}) \,, \tag{6}$$

whereas if the antenna separation is less than the inner scale, the visibility will be Gaussian. By noting a change in the form of the visibility function from the form equation (6) to a Gaussian as the baseline lengths decrease, one can measure the inner scale. This procedure was used by Spangler and Gwinn (1990) on published measurements of α, to conclude that the inner scale in the interstellar medium is $50-200$ kilometers. The value of determining the inner scale is that it is plausibly identified with a plasma microscale in the medium, which in turn allows physical properties of the medium to be identified. Spangler and Gwinn concluded that the medium hosting the irregularities has properties characteristic of the diffuse envelopes of H II regions and not the coronal phase of the interstellar medium.

The results of Spangler and Gwinn (1990) referred to the highly scattering or "Type B" component of interstellar turbulence (Cordes, Weisberg, and Boriakoff 1985). The simple reason for this is that for the scattering to be detectable and accurately measured with interferometers operating at frequencies of hundreds or thousands of MHz, the scattering must be rather heavy. Cordes, Weisberg, and Boriakoff (1985) noted that there is another, more diffuse "Type A," scattering medium as well. The properties of this medium are not as well specified, and the inner scale cannot be determined in the same way. The reason for this is the coefficient A in equation (5) is proportional to the square of the wavelength and the scattering measure. For path lengths which pass only through the "Type A" medium, and for interferometer baselines small enough to resolve the inner scale, the scattering disk is unresolved at frequencies of hundreds or thousands of MHz. This situation can only be resolved by making observations at lower frequencies. The benefits of such observations would be an identification of the plasma characteristics in the medium supporting the Type A scattering material, which would contribute greatly to an understanding of hydromagnetic wave properties in this part of the interstellar medium.

Low-Frequency Variability and Refractive Scintillation. Since the work of Rickett, Coles, and Bourgois (1984), it has been widely accepted that the low-frequency variability of extragalactic radio sources is due to the refractive interstellar scintillation. However, while this theory is capable in general terms of accounting for the observed variations, a detailed comparison of the model with data reveals discrepancies (Spangler *et al.* 1989). Further work is needed, both to verify the scintillation nature of low-frequency variability, and also to extract the information about the interstellar medium contained by these variations.

A crucial feature of the scintillation model is that the "diffractive angular size," *i.e.*, the scattering size which would be measured with an interferometer for a point source, be comparable to the intrinsic size of the source (Rickett 1986). If the scattering size is much larger than the intrinsic size, strong scintillations will result. For the relatively weak variations which characterize typical low-frequency variability (scintillation indices of 2–4% are the norm (Spangler *et al.* 1989), the diffractive scattering size can be substantially less than the intrinsic size, but not altogether negligible. Given this basic feature of the refractive scintillation model, one would expect at sufficiently low frequencies the diffractive size should dominate, for reasons mentioned in the Introduction in connection with equation (1). From low-frequency measurements of the sources, one could determine the diffractive angular size at the decimetric wavelengths at which flux variations are observed and check for consistency. Such an analysis would constitute a definitive test of the refractive scintillation model.

5   Conclusions

A low-frequency interferometer in space is potentially an important tool for the study of diffuse plasma turbulence in the interstellar medium. The main points presented in this paper are as follows.

(1)   The subject of radio wave scattering in the interplanetary medium needs to be thoroughly addressed in the early stages of planning for this instrument. If the instrument is sufficiently insensitive and requires long integration times, so the ensemble average visibility is measured, the instrument will be *less* sensitive to diffuse turbulence than existing VLBI interferometers.

(2)   One of the more interesting possible applications of this instrument might be the detection of density perturbations due to hydromagnetic waves in the vicinity of supernova remnants and other interstellar shock waves. This might permit a direct test of the Fermi mechanisms for particle acceleration in these objects.

(3)   A low-frequency interferometer would be well suited for additional measurements relevant to the diffuse phase of interstellar turbulence. Examples of observations which could be made are a measurement of the inner scale of the plasma turbulence and a definitive test of the refractive scintillation hypothesis for low-frequency variability.

References

Armstrong, J.W., Woo, R., 1980. Jet Propulsion Interoffice Memorandum 3331-80-070.

Armstrong, J.W., Woo, R., 1981. Astr. Astrophys., **103**, 415.

Bourgois, G., Cheynet, C., 1972. Astr. Astrophys., **21**, 33.

Coles, W.A., 1978. Space Sci. Rev., **21**, 411.

Coles, W.A., Harmon, J.K., 1989. Astrophys. J., **337**, 1023.

Cordes, J.M., Weisberg, J.M., Boriakoff, V., 1985. Astrophys. J., **288**, 221.

Cordes, J.M., Rickett, B.J., Backer, D.C., 1988. In *Amer. Inst. Phys. Conf. Proc. #174, Radio Wave Scattering in the Interstellar Medium.*

Cornwell, T.J., Anantharamaiah, K.R., Narayan, R., 1988. In *Amer. Inst. Phys. Conf. Proc. #174, Radio Wave Scattering in the Interstellar Medium*, p. 92.

Erickson, W.C., 1964. Astrophys. J., **139**, 1290.

Goodman, J., Narayan, R., 1989. Mon. Not. R. Astr. Soc., **238**, 995.

Gwinn, C.R., Moran, J.M., and Reid, M.J., 1988. In *Amer. Inst. Phys. Conf. Proc. #174, Radio Wave Scattering in the Interstellar Medium*, p. 106.

Lee, M.A., 1982. J. Geophys. Res., **87**, 5063.

Marscher, A.P., 1977. Astrophys. J., **216**, 244.

Mutel, R.L., Hodges, M.W., 1986. Astrophys. J., **307**, 472.

Rickett, B.J., 1977. Ann. Rev. Astr. Astrophys., **15**, 479.

Rickett, B.J., Coles, W.A., Bourgois, G. 1984. Astron. Astrophys., **134**, 390.

Spangler, S.R., Mutel, R.L., Benson, J.M., Cordes, J.M., 1986. Astrophys. J., **301**, 312.

Spangler, S.R., Cordes, J.M., 1988. Astrophys. J., **322**, 346.

Spangler, S.R., Fey, A.L., Cordes, J.M., 1987. Astrophys. J., **322**, 909.

Spangler, S.R., Fanti, R., Gregorini, L., Padrielli, L., 1989. Astron. Astrophys. J., **209**, 315.

Spangler, S.R., Gwinn, C.R., 1990. Astrophys. J. (in press).

LOW FREQUENCY INTERSTELLAR SCATTERING AND PULSAR OBSERVATIONS

James M. Cordes
Astronomy Department, Cornell University

Radio astronomy at frequencies from 2 to 30 MHz challenges time tested methods for extracting usable information from observations. One fundamental reason for this is that propagation effects due to the magnetoionic ionosphere, interplanetary medium, and interstellar medium (ISM) increase strongly with wavelength. In this paper I address the problems associated with interstellar scattering off of small scale ($\sim 10^9$ cm) irregularities in the electron density.

I will first summarize what we know of interstellar scattering on the basis of high frequency observations, including scintillation and temporal broadening of pulsars and angular broadening of various galactic and extragalactic radio sources. Then I will address those high frequency phenomena that are important or, at least, detectable at low frequencies. The radio sky becomes much simpler at low frequencies: most pulsars will not be seen as time varying sources, intensity variations will be quenched or will occur on time scales much longer than a human lifetime, and many sources will be angularly broadened and/or absorbed into the noise. Angular broadening measurements will help delineate the galactic distribution and power spectrum of small scale electron density irregularities. Images of scattered sources can take on the form of the distribution of the scattering material in this low frequency regime. Spectral broadening may be a relevant line broadening mechanism for low frequency recombination lines.

Summary of Interstellar Scattering and Scintillation Phenomena

Interstellar scattering phenomena are rich in variety. *Diffraction* from small scale irregularities is manifest as angular broadening (interstellar 'seeing'); temporal broadening of pulsar pulses due to multipath propagation; and diffractive intensity scintillations. The latter have been seen only from pulsars owing to their small size (the 'stars twinkle, planets do not' rule: the critical angular size for interstellar scintillation is typically less than 1 *micro* arc sec.) Another effect, as yet unobserved from the ISM, is spectral broadening, the smearing of a spectral line due to modulations imposed by the diffracting medium. *Refraction* effects from larger irregularities include long term scintillations; wandering of apparent source positions on the sky (angle of arrival [AOA] variations); and time of arrival variations associated with the AOA variations and also with the changing electron column density (the dispersion measure, DM) as turbulence is swept across the line of sight. Figure 1 shows examples of several of these phenomena, including the visibility function of a scattered pulsar; temporal broadening of a pulsar pulse; diffractive intensity scintillations; and dispersion measure variations.

There is substantial evidence that the electron density variations responsible for the phenomena span a large range of length scale and that they may be described by a

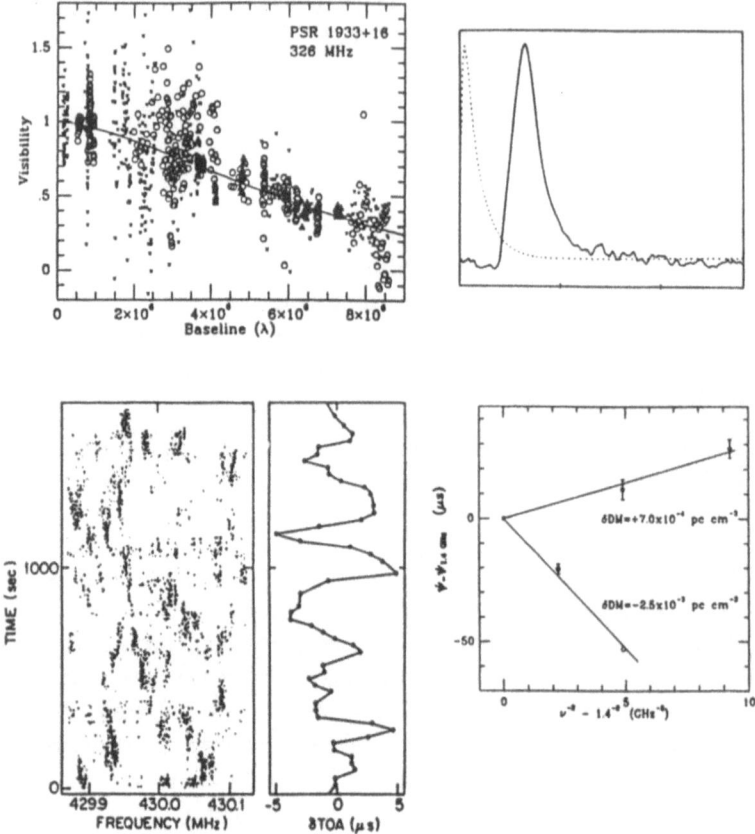

Figure 1. Examples of scattering observables: (a) visibility function[1]; (b) pulsar pulse with scattering broadening tail[2]; (c) grey scale plot of diffractive intensity scintillations[3]; (d) dispersion measure variations[4].

spatial power spectrum that is power law in form. Thus we define the power spectrum in terms of the mean square electron density:

$$\langle \delta n_e^2 \rangle \equiv \int_{q_0}^{q_1} d^3q \, C_n^2 \, q^{-\alpha}, \tag{1}$$

where we introduce the coefficient C_n^2, wavenumber cutoffs $q_{1,2}$, and spectral index α. Pulsar scintillations and time of arrival variations, and limits on AOA variations of OH masers[5] imply that $3.5 \leq \alpha \leq 4.0$. The Kolmogorov value, $\alpha = 11/3$, is consistent with many lines of sight. Defining, respectively, the 'inner' and 'outer' length scales $\ell_1 = 2\pi/q_1$ and $\ell_0 = 2\pi/q_0$ of the spectrum, limits from several kinds of observations suggest that $\ell_1 \leq 10^9$ cm and $\ell_0 \geq 10^{14}$ cm. Recent work[5,6,7] suggests that the inner scale is of the order of 100 km for a few heavily scattered lines of sight, a length scale consistent with the gyro radius of thermal ions in the warm (10^4 K) phase of the ISM. There are indirect arguments that the outer scale extends to parsec scales, based on studies of rotation measure variations[8] and on the notion that cosmic ray diffusion *requires* magnetic irregularities on all scales from about 1 AU to 1 pc and that these are accompanied by density irregularities[9].

The easiest quantity to estimate, given the spectral index α, which we now assume to be 11/3, is the line of sight integral of C_n^2, which we call the *scattering measure*. The scattering measure and estimates of it using angular scattering diameters of extragalactic sources θ_{FWHM} and pulsar temporal broadening times τ_d are:

$$SM \equiv \int_0^D ds\, C_n^2(s) = \left(\frac{\theta_{FWHM}}{\theta_0}\right)^{5/3} \nu^{11/3} = A_\tau \left(\frac{\tau_d}{D}\right)^{5/6} \nu^{11/3},$$

where $\theta_0 = 0.13$ arc sec, $A_\tau = 387$ kpc m$^{-20/3}$ for τ_d in seconds and D in kpc, and ν in GHz. Measured values of SM range from 10^{-4} to 10^3 kpc m$^{-20/3}$. For scintillation bandwidth measurements, SM may be estimated by first using the 'uncertainty' relationship $2\pi \Delta\nu_d \tau_d = 1$ to estimate the effective temporal broadening time. For a pulsar embedded in a homogenously turbulent medium, we have

$$\theta_{FWHM} = \left[\frac{16(\ell n2)c\tau_d}{\pi D}\right]^{1/2}.$$

In terms of the scattering diameter, the diffraction scale is $\ell_d = 2\sqrt{\ell n2}\lambda/\pi\theta_{FWHM}$.

Spectral broadening is approximately equal to the reciprocal of the intensity scintillation time; both are due to modulations of the signal due to motion of diffracting material across the line of sight. In terms of SM, the broadening is[10]

$$\Delta\nu_s \approx \frac{V_\perp}{\ell_d} \approx 2\pi V_\perp [4\pi r_e^2 \lambda^2 C_n^2 D]^{3/5} = 1.16 \text{ Hz } V_{100}\nu^{-6/5}SM^{3/5},$$

where r_e is the classical electron radius, $V_\perp$ is the relevant transverse speed (a combination of pulsar, ISM, and Earth velocities), V_{100} is in units of 100 km s^{-1}, and SM has units of *kpc m$^{-20/3}$*, and ν is in GHz.

For galactic objects with known distances, we obtain the line of sight average $\overline{C_n^2} \equiv SM/D$. The remarkable fact is that $\overline{C_n^2}$ varies by a factor of at least 10^4 between different lines of sight. A homogenously turbulent medium would, of course, yield a constant value for this quantity. Figure 2 shows SM plotted against galactic latitude and longitude on an equal area projection. The plotted circles have diameters that are proportional to $\log SM$. It is obvious that scattering is a galactic phenomenon, since the largest circles are on lines of sight in the galactic plane. Study of how SM varies (work in progress) implies that the variation in $\overline{C_n^2}$ is due to both large scale galactic structure and localized regions of intense turbulence, including HII regions and supernova remnants. Enhanced scattering seems to be correlated over small angular scales, implying that the depth is of order a few parsecs or less. This implies that the internal C_n^2 of enhanced regions is correspondingly larger than $\overline{C_n^2}$ by a factor of about 1000.

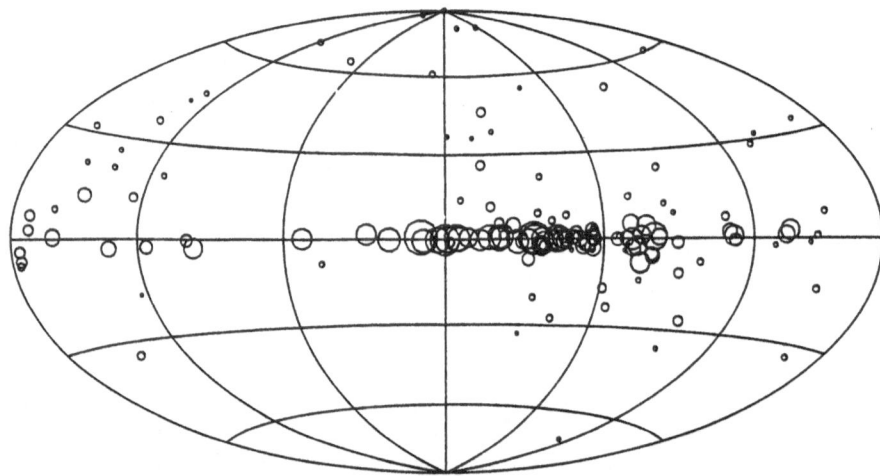

Figure 2. Scattering measures plotted versus ℓ, b; $\log SM = 3$ for the largest circles, the smallest circles have $\log SM = -4$.

To make estimates of low frequency scattering phenomena, it is useful to fit an idealized model to the data. The *local* value of C_n^2 at galactocentric radius R and distance z from the galactic plane comprises two components:

$$C_n^2(R, z) = C_1 \exp[-(R^2/A_1^2 + z/H_1)] + C_2 \exp[-(R^2/A_2^2 + z/H_2)].$$

A grid search to minimize the mean square difference between predicted and measured SM yields a large scale component with $A_1 \gtrsim 10$ kpc and $H_1 \approx 1$ kpc and a galactic center component with $A_2 \approx 3.0$ kpc and $H_2 \approx 0.05$ kpc. The strengths of the two components are $C_1 \approx 10^{-4}$ m$^{-20/3}$ and $C_2 \approx 5$ m$^{-20/3}$. This model accounts for the broad brush structure that determines SM, but there are substantial departures from the model's predictions. Figure 3 shows the modeled SM plotted against measured SM. Notable lines of sight that deviate from the model by more than two orders of magnitude include the Vela pulsar[11]; Cygnus X-3 (ref 6); and the extragalactic source behind the HII complex NGC6334B (ref 5). The grid search was iterated several times so that 'deviant' sources could be recognized and excluded from the fit. The numbers quoted above derive from the end result of this process. In a later section, the model is used to predict observable quantities such as angular broadening, spectral broadening, and temporal broadening.

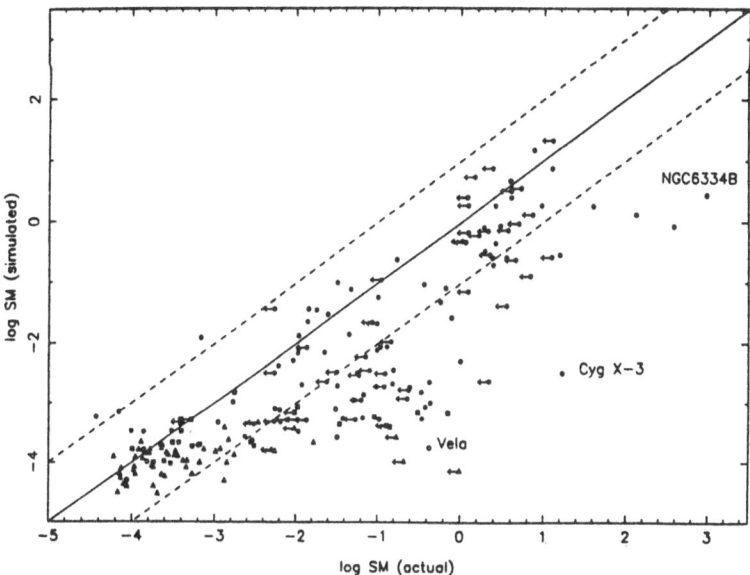

Figure 3. Modeled SM plotted against actual SM. A few notable lines of sight are labeled.

TABLE 1
DIFFRACTION PHENOMENA

Phenomenon	Quantity	Typical Values		
		400 MHz	30 MHz	10 MHz
Intensity Scintillations	$\Delta\nu \propto \lambda^{-4.4}$	10 kHz	10^{-2}	10^{-7}
	$\Delta t_d \propto \lambda^{-1.2}$	60 sec	3	0.7
Temporal broadening	$\tau_d \propto \lambda^{4.4}$	1 ms	1 min	3 hr
Angular broadening	$\theta_d \propto \lambda^{2.2}$	10 $m.a.s.$	$3''$	$33''$
Spectral broadening	$\Delta\nu_s \propto \lambda^{1.2}$	1 Hz	22	83

TABLE 2
REFRACTION PHENOMENA

Phenomenon	Quantity	Typical Values		
		400 MHz	30 MHz	10 MHz
Slow scintillations	$\sigma_I/I \propto \lambda^{-1.1}$	20%	1	0.3
	$\Delta t_r \propto \lambda^{2.2}$	1 yr	300	3000
Time-of-arrival variations	$\Delta t_{DM} \propto \lambda^2$	$10\mu s$	2 ms	16 ms
	$\Delta t_\theta \propto \lambda^{1.6}$	$1\mu s$	60	400
	$\Delta t_{\theta^2} \propto \lambda^{3.3}$	$1\mu s$	5 ms	200 ms
Angular wandering	$\Delta \theta_r \propto \lambda^{1.6}$	1 $m.a.s.$	60	400

Interstellar Scattering at Low Frequencies

Using results from high frequency observations and our understanding of the scattering medium's spectrum and distribution, it is possible to discuss the phenomena at low frequencies. In Table 1 the scaling laws[12] for various diffraction phenomena are presented and representative values of the observables are given. Intensity scintillations, which have typical frequency and time scales of 10 kHz and 60 sec at meter wavelengths become sub-Hz and subsecond at 10 MHz. In order to detect intensity scintillations, a radiometer must have sufficient signal to noise ratio for the frequency and time resolution necessary to resolve the scintillations in both time and frequency. At low frequencies where system temperatures are determined by the sky background, it is clear that diffractive scintillations will never be detectable, even for the nearest pulsar, which has the largest scintillation bandwidth. Temporal broadening becomes much larger than the pulse period of almost all pulsars at 10 MHz, so pulsars will be seen as steady sources. Angular broadening becomes a few to many arc seconds. Spectral broadening may be important for recombination line observations.

Figure 4 details the observed scattering broadening times of pulsars, showing τ_d plotted against DM. Data are taken from the literature and have been scaled to 1 GHz assuming the Kolmogorov scaling $\tau_d \propto \lambda^{22/5}$. To assess the level of temporal broadening at low frequencies, horizontal lines have been drawn (again using the Kolmogorov scaling) to show 1 sec of broadening at 10 and 30 MHz. The meaning of these lines is that all pulsars above each horizontal line will be essentially smeared out, since the average pulsar period is $\sim$ 0.7 sec. At 10 MHz, there are only 2 pulsars below the line, while at

30 MHz, a few tens of objects are below the line. At 2 MHz, all pulsars are well above the 1 sec line (not shown). It is amusing to consider the most broadened pulsar (PSR 1849+00; ref 13) which has 1 sec of broadening even at 1 GHz. The formal scaling to 1 MHz, yields a broadening of 0.5 Myr! Note, however, that the scaling law $\tau_d \propto \lambda^{22/5}$ is based on a small angle approximation[14], which breaks down for heavily scattered lines of sight at low frequencies. Relaxing the small angle approximation gives a more reasonable result: only about 20 kyr! This means that the distribution of paths along which radiation propagates subtends a major fraction of the galactic disk. Of course, this is an academic exercise, since no source will be observable along such heavily scattered lines of sight due to free free absorption (see below).

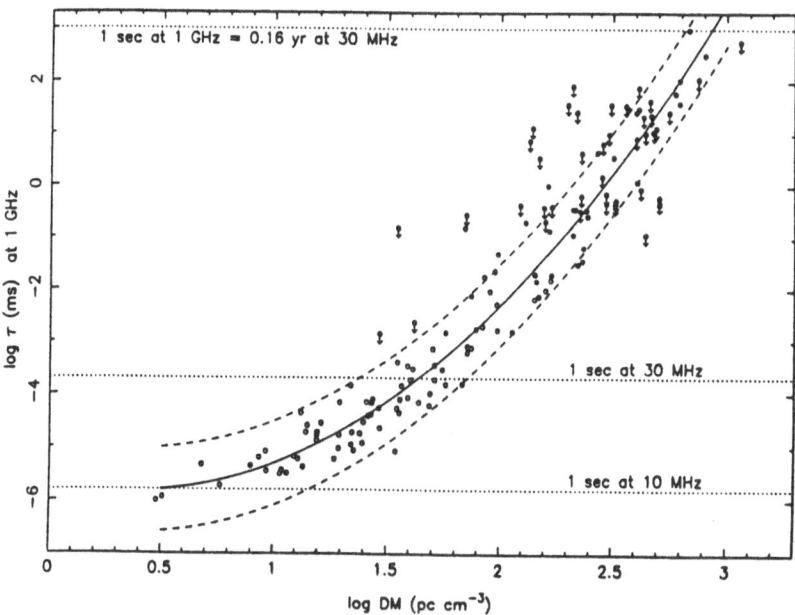

Figure 4. Pulsar temporal broadening times plotted against dispersion measure, DM. The solid line is a least squares fit to the data; the dashed lines are $\pm 1\sigma$ deviations from the fit. Downward arrows denote upper limits, which were excluded from the fit.

Refraction phenomena are presented in Table 2. Refractive scintillations occur on time scales measured in hundreds of years, so even the most dedicated of monitorers will be thwarted in variability studies. Angle of arrival variations and the associated time of arrival variations scale more slowly with wavelength than do angular and temporal broadening. As a consequence, these variations will be proportionally smaller than the angular size or pulse width and thus extremely difficult to measure at low frequencies.

Predicted Observables at Low Frequencies

The net implications for low frequency radio astronomy are that: (1) Diffractive intensity scintillations will be quenched for realistic instrumental bandwidths and time constants. This means that the equivalent of speckle interferometry in the radio will not be possible in imaging observations. (2) Pulsar periodicities will be quenched by temporal broadening. This statement holds for all pulsars at 2 MHz, all but a few at 10 MHz, and for most at 30 MHz. (3) The time scale of refractive phenomena is much longer than a scientific career. The radio sky indeed will appear quiesent compared to high frequencies. (4) The dominant observable will therefore be the angular broadening of all kinds of radio sources. For a few pulsars in the 10 to 30 MHz band, temporal broadening can be measured, leading to determinations of the scaling law of the temporal broadening, in turn leading to further constraints on the slope and cutoffs of the wavenumber spectrum. Also, for a few pulsars, it will be possible to test whether time of arrival variations that depend on wavelength follow the λ^2 dependence of the cold plasma dispersion law, or whether TOA variations due to AOA variations are also important. These last tests will help constrain the wavenumber spectrum for δn_e on scales much larger than 1 AU. (5) Sources viewed at low galactic latitudes will be so scattered that their shapes will take on the form of the distribution of scattering material, viz. the galactic plane. This is because the maximum traversal of a ray transverse to the propagation direction is limited by the extent of the scattering medium. This source of image asymmetry has not been addressed before and competes with asymmetries due to anisotropies of the small scale irregularities themselves and distortion of a scattered image by refraction[15,16].

Using the idealized model for $C_n^2(R,z)$ discussed above, it is possible to predict the values of the most important observables at low frequencies, namely the angular broadening diameter θ_{FWHM} and the spectral broadening $\Delta\nu_c$. Figure 5 shows the angular broadening plotted against galactic latitude for several cuts in longitude.

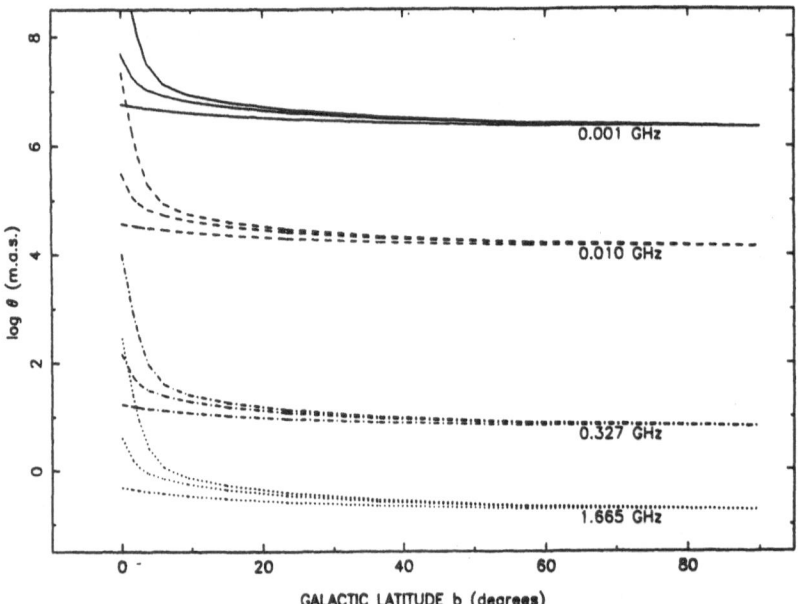

Figure 5. Predicted angular broadening of extragalactic sources vs. galactic latitude for different frequencies and for three longitudes (0, 60, and 180°.)

Spectral broadening takes on values of 300, 20, and 5 kHz at observing frequencies of 1, 10, and 30 MHz for the largest observed scattering measure (10^3 kpc $m^{-20/3}$ for NGC6334B, ref 5). This broadening corresponds to velocity spreads of $10^5, 500,$ and 50 km s^{-1}. At high latitudes, the spectral broadening is a factor of 10^4 smaller and is therefore negligible.

The spectrum defined in equation (1) implies a *lower bound* on the emission measure for a given line of sight. The total emission measure is:

$$EM = \int_0^D ds \, \langle n_e^2 \rangle = \int_0^D ds \, [\langle n_e \rangle^2 + \langle (\delta n_e)^2 \rangle] \equiv EM_0 + EM_\delta.$$

The line of sight integral of equation (1) gives the second term on the right hand side of the above equation. The integral of eqn (1) is

$$\langle (\delta n_e)^2 \rangle = \frac{4\pi (2\pi)^{3-\alpha}}{\alpha - 3} C_n^2 \ell_0^{\alpha-3} \left[1 - \left(\frac{\ell_1}{\ell_0} \right)^{\alpha-3} \right] \approx \frac{4\pi}{\alpha - 3} C_n^2 \left(\frac{\ell_0}{2\pi} \right)^{\alpha-1}.$$

for $\alpha \neq 3$, where the approximate equality holds for the inner scale being much smaller than the outer scale, $\ell_1 \ll \ell_0$, which seems to be a good approximation. For $\alpha = 11/3$ we have

$$EM_\delta \approx 544 \, \text{pc cm}^{-6} \, SM(\text{kpc m}^{-20/3}) \, [\ell_0(\text{pc})]^{2/3}.$$

The measured values of SM therefore imply minimum emission measures of 0.05 to 10^5 pc cm^{-6} if we assume an outer scale of 1 pc. In Figure 6 we show EM_δ plotted against ℓ, b using an outer scale of 100 pc as would be measured along the lines of sight to extragalactic sources. Lines are also drawn that designate optical depth unity for several radio frequencies. It is clear that much of the Galaxy is opaque at these low frequencies. It is of interest to compare the predicted minimum EM with measured values of Reynolds[17]. For $|b| = 90°$, Reynolds obtains $EM \approx 4$ pc cm^{-6} while the scattering measurements suggest $EM_\delta(|b| = 90°) \approx 0.11 \ell_0^{2/3}$. If we assume that $EM_0 \approx EM_\delta$, then an outer scale of $\ell_0 \approx 100$ pc is implied. Such an outer scale is consistent with other conjectures about the overall extent in wavenumber of the electron density spectrum[18,19].

References

1. Gwinn, C.R., Cordes, J.M., Bartel, N., and Wolszczan, A. 1988, *Ap. J. (Letters)*, **334**, L13.

2. Cordes, J.M., Weisberg, and Boriakoff, V. 1985; *Ap. J.*, **288**, 221.

3. Cordes, J.M., Wolszczan, A., Dewey, R. J., Blaskiewicz, M., and Stinebring, D. R. 1990; *Ap. J.*, **349**, 245. See also Rawley, L.A., Taylor, J. H., and Davis, M.M. 1988, *Ap. J.*, **326**, 947.

4. Gwinn, C.R., Moran, J. M., Reid, M. J., Schneps, M. H. 1988, *Ap. J.*, **330**, 817.

5. Moran, J.M., Rodriquez, L. F., Greene, B. and Backer, D.C. 1989, *Ap. J.*, in press.

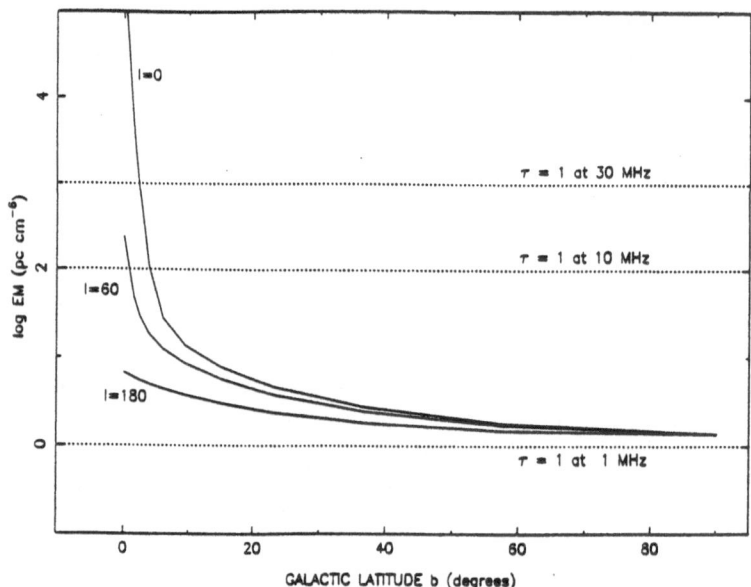

Figure 6. Predicted minimum emission measure vs. galactic coordinates.

6. Molnar, L. A., Mutel, R. L., Reid, M. J., and Johnston, K. J. 1990, *Ap. J.*, submitted.

7. Spangler, S.R. and Gwinn, C.R. 1990, *Ap. J.*, in press.

8. Simonetti, J. H., Cordes, J.M., and Spangler, S.R. 1984, *Ap. J.*, **284**, 126; Lazio, J., Spangler, S.R., and Cordes, J.M., *Ap.J.*, in press.

9. Jokipii, J. R. 1988, in AIP Proceedings 174, *Radio Wave Scattering in the Interstellar Medium*, 48.

10. Lazio, J. and Cordes, J.M., in preparation.

11. Backer, D. C. 1974; *Ap. J.*, **190**, 667.

12. Cordes, J. M., Pidwerbetsky, A., and Lovelace, R.V. 1986, *Ap. J.*, **310**, 737. See also Romani, R., Narayan, R., and Blandford, R. D. 1986, *M.N.R.A.S.*, **220**, 19; Foster, R. S. and Cordes, J. M. 1990, *Ap. J.*, submitted.

13. Clifton, T. 1985, Ph.D. Thesis, Univ. of Manchester.

14. Rickett, B. J. 1977, *Ann. Rev. Ast. Ap.*, **15**, 479.

15. Spangler, S. R. and Cordes, J. M. 1988, *Ap. J.*, **332**, 346.

16. Narayan, R. and Goodman, J. 1989, *M.N.R.A.S.*, **238**, 963.

17. Reynolds, R. J. 1990, these proceedings.

18. Armstrong, J. W., Cordes, J. M., and Rickett, B. J. 1981, *Nature*, **291**, 561.

19. Lee, L.C. and Jokipii, J. R. 1976, *Ap. J.*, **206**, 735.

Pulsar Astronomy at Meter and Decameter Wavelengths: Results from Arecibo

J.A. Phillips and A. Wolszczan
National Astronomy and Ionosphere Center
Cornell University

Abstract

We review the state of groundbased research in pulsar astronomy at low frequencies with an emphasis on new results from Arecibo. We have obtained high signal-to-noise profiles for six pulsars at 25, 47 and 111.5 MHz, along with several higher frequencies. Three of the pulsars in our sample had been previously reported to exhibit sporadic and intense interpulse emissions at decameter wavelengths. Our 25 MHz data show no evidence of off-pulse activity exceeding 4% of the main pulse intensity for PSR 0834+06, 2% for PSR 0950+08, and 1% for PSR 1133+16. Multifrequency timing data collected over a broad frequency range (25 → 4800 MHz) were used to derive accurate pulsar dispersion measures. We find that pulse arrival times obeyed a cold plasma dispersion law to high accuracy over a 200:1 frequency range. This result is compared with reports of low-frequency "superdispersion." The timing data have also been used to set limits on the size and location of the radio emitting region in six pulsars. By analyzing the effects of retardation, abberation and magnetic sweepback on pulse arrival times we find the radial extent of the emission region to be < 200 km over a 100:1 frequency range. The data indicate emission altitudes less than 3 stellar radii above the surface of the neutron star.

1. Introduction

Pulsar observations at decameter wavelengths are well suited for studies of astrophyical plasmas both in the interstellar medium (ISM) and in neutron star magnetospheres. Radio propagation effects in plasmas are usually characterized by a steep frequency scaling. Cold plasma dispersion in the ISM, for example, introduces a group delay $\tau_g \propto DM\nu^{-2}$ into pulsar arrival times. At decameter wavelengths, the interstellar electron column density (DM) can be measured from the dispersion of pulsar signals with an accuracy exceeding 1 part in 10^5. Accurate DM measurements can, in turn, be used to study the spatial spectrum of interstellar turbulence (Rickett 1988), to constrain physical conditions in the interstellar plasma (Tanenbaum et al. 1968), and to measure the size of pulsar emission regions (Cordes 1978). Low frequency pulsar observations may also be useful as a probe of magneto-ionic conditions in the pulsar magnetosphere, especially near the light cylinder where the observing frequency could be comparable to the gyro- and plasma frequencies.

The high expectations for low frequency pulsar research have perhaps been heightened further by recent reports that:

1) pulse broadening by interstellar scattering is more than an order of magnitude smaller at meter wavelengths than predicted by theory (Kuz'min et al. 1988);

2) pulsar profiles at decameter wavelengths exhibit intense and sporadic steep spectrum components at longitudes outside the main pulse (e.g., Bruck 1987); and

3) pulsar dispersion measures increase with decreasing radio frequency (Kuz'min 1987; Shitov et al. 1988).

Following nomenclature used in the literature we will refer to the latter two effects as "low-frequency interpulse emission" (IPE) and "superdispersion," respectively. In an effort to investigate low frequency pulsar emission, we have initiated a program to observe all the low DM pulsars visible from Arecibo using receiving systems at 25 MHz, 47 MHz and 112 MHz, along with several higher frequencies. In this paper we present our observations together with some initial analyses of the results.

The paper is divided into two main areas: pulse morphology (§2) and multi-frequency timing (§3). In §2 we present high signal-to-noise profiles for six pulsars at 25 MHz and discuss the evidence for and against low-frequency IPE. §3 contains the results of a timing program to measure pulsar DMs over a broad frequency range (25 → 4800 MHz). The DM results are used in §3.3 for a critical analysis of superdispersion and in §3.4 to constrain emission radii in the pulsar magnetosphere.

2. Pulse Morphology

Figure 1 displays the 25 MHz pulse profiles of six pulsars observed at the Arecibo Observatory. Due to high levels of man-made interference at decameter wavelengths, the best times to observe are during long winter nights when the terrestrial ionosphere is thin. Consequently, we have concentrated our efforts on pulsars which transit at night between November and March (i.e., in the 7 to 16 hr sidereal range). The data presented here were collected during the winters of 1988 and 1989. For a complete description of our observing procedures, see Phillips and Wolszczan (1989).

The profiles in Figure 1 are, in general, very much like profiles for the same pulsars at higher frequencies. This result is in apparent disagreement with the observations of Bruck and Ustimenko (1976, 1977, 1979) who report intense steep-spectrum emission outside the main pulse in profiles below ∼ 50 MHz. In Figure 1, the strongest limits on off-pulse emission are set by PSR 0834+06, PSR 0950+08, and PSR 1133+16. No IPE is evident above the following 2σ noise levels: 4% of the main pulse height for PSR 0834+06, 2% for PSR 0950+08, and 1% for PSR 1133+16. These results improve on the limits previously set by Phillips and Wolszczan (1989). For the same pulsars Bruck and Ustimenko (1976, 1979) and Bruck (1987, 1989) find interpulses as strong as 30% to 100% of the main pulse using integration times from 10 minutes to 3 hours with 10 to 100 kHz bandwidths.

To test for the possibility of narrowband, sporadic IPE we divided our data into 78 kHz frequency bins and examined individual 10 min integrations. Our total observing bandwidth for PSR 0950+08 and PSR 1133+16 (PSR 0834+06) was 625 kHz (312.5 kHz) with 1.22 kHz (0.61 kHz) frequency resolution. Typical 2σ noise levels were 15-20% of the main pulse; no off-pulse emission was evident above that level.

We still cannot rule out the existence of weak IPE which fluctuates on time scales less than 10 min with bandwidths much smaller than 78 kHz. Indeed, the possibility of such emission has recently been raised by Reyes (1989) who reports transient IPE in

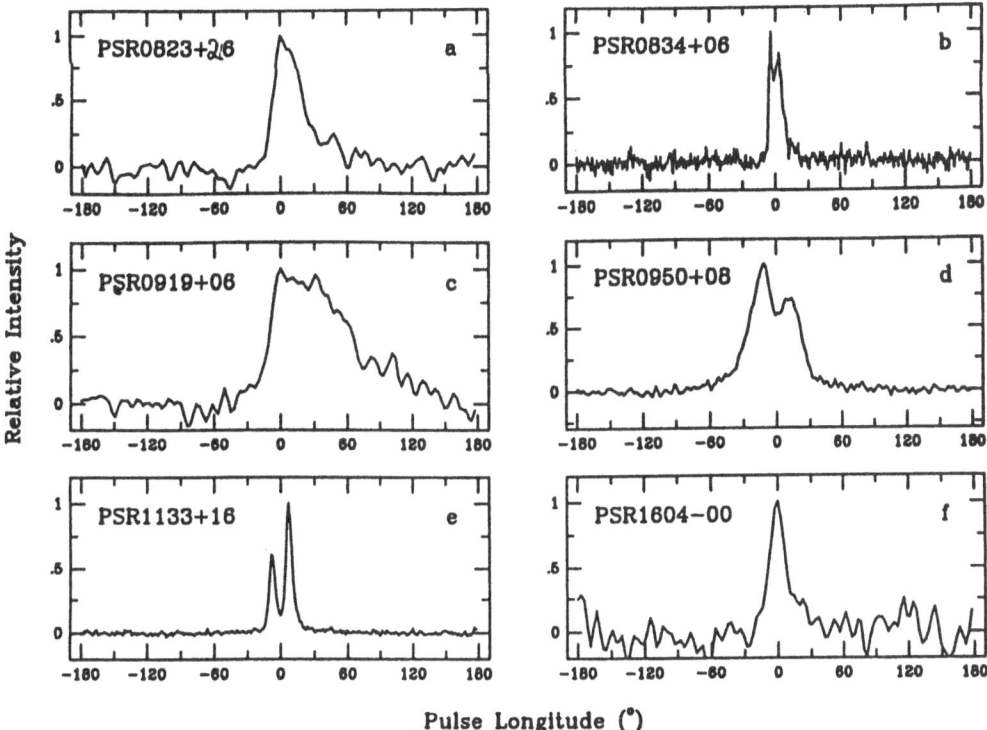

Figure 1: 25 MHz pulse profiles collected at the Arecibo Observatory. Integration times were: a) 6.0 hr, b) 4.0 hr, c) 4.5 hr, d) 8.3 hr, e) 11.7 hr and f) 2.8 hr.

single pulses from PSR 0834+06 at 45 MHz using a 10 kHz observing bandwidth. Another possibility is that IPE, as reported by Bruck and Ustimenko, is real but extremely rare. It is notable, however, that we have observed PSR 1133+16 for ~ 700 minutes with no detection of off-pulse emission. Bruck (1989) recently reported the detection of a sporadic burst of IPE simultaneously at the Pushchino Observatory near Moscow and at the Khar'kov Observatory in the Ukraine. This may prove to be an important observation but, for the moment, one must remain skeptical of results obtained with short integration times and narrow bandwidths since those kind of data are inherently noisy and can easily lead to the spurious detection of off-pulse features.

3. Multi-frequency Timing

Pulsar signals are dispersed by ionized gas as they propagate through the interstellar medium. The arrival time delay between two frequencies is given by

$$t(\nu_2) - t(\nu_1) = 4.15 \times 10^9 DM(\nu_2^{-2} - \nu_1^{-2}) \qquad (1)$$

with t in μs, ν in MHz, and DM in pc cm^{-3}. The measurement of pulsar DMs has recently excited much interest, largely because of persistent reports that dispersion measures increase rapidly at low frequencies (e.g., Shitov et al. 1988). To investigate the frequency dependence of pulsar dispersion, we have implemented a multi-frequency timing program at Arecibo to obtain accurate DMs for six nearby pulsars, four of which were previously reported to exhibit "superdispersion."

3.1 Observations

Between June and December 1989 we used the Arecibo 305m radiotelescope to observe six pulsars: PSR 0823+26, PSR 0834+06, PSR 0919+06, PSR 0950+08, PSR 1133+16, and PSR 1604-00. The observing frequencies (feeds) were 25, 47 and 111.5 MHz (dual circularly polarized crossed-dipole point feeds); 318 MHz (a linearly polarized line feed); 430, 1408 and 2380 MHz (dual circularly polarized line feeds); and 4800 MHz (a Gregorian-type subreflector).

A three-level, 40 MHz autocorrelation spectrometer was used to integrate 64-lag to 512-lag autocorrelation functions (ACFs) of the input voltage into 128 to 2048 pulse phase bins synchronously with the Doppler-corrected pulse period. ACFs were written to magnetic tape during the observations and Fourier transformed off-line to obtain power spectra as a function of pulse phase. The time resolution (# of phase bins) and frequency resolution (# of lags) were set depending on the dispersion measure of the pulsar and the observing frequency. Observing bandwidths were 312.5 or 625 kHz at 25, 47 and 111.5 MHz, 10 MHz at 318 and 430 MHz, and 40 MHz at the higher frequencies. Effective time resolutions were typically 5 ms at 25 MHz, 1.5 ms at 47 MHz, 1 ms at 111.5 MHz and 0.5 ms between 1408 and 4800 MHz.

The integration time for each ACF was 2 to 10 minutes. Local time at the beginning of each integration was recorded to ± 1 μs using a rubidium clock. The rubidium standard was referenced to the Loran C time service.

During a typical observing session, each pulsar was observed at 430 MHz plus one or more other frequencies. The measurements were repeated at least twice (i.e., on two different nights) to verify the repeatability of timing results. At 25 and 47 MHz we observed each pulsar 5 to 15 times over at least one month to check for DM variability and to build up a high signal-to-noise profile.

To assess the effect of the solar corona and the interplanetary medium on our measurements, we observed four pulsars (PSR 0823+26, PSR 0834+06, PSR 0919+06, and PSR 0950+08) at 47 and 430 MHz over a wide range of solar elongation ($e = 5° \rightarrow 120°$). We found the dispersion measure to be independent of solar elongation for $e > 10°$ within an uncertainty of 2×10^{-4} pc cm^{-3}. Between 5° and 10° elongation, the corona contributed as much as 15×10^{-4} pc cm^{-3} to the DM. We therefore discarded all data with $e < 10°$.

3.2 Results

Figure 2 displays the multi-frequency pulse profiles. To measure delays in arrival time between different frequencies, it was necessary to identify a fiducial point which corresponded to the same rotational phase in each profile. For pulsars with single profiles like PSR 0823+26 we selected the pulse maximum as the fiducial point. For double profiles we used the midpoint between peaks (Craft 1970). In some cases, like

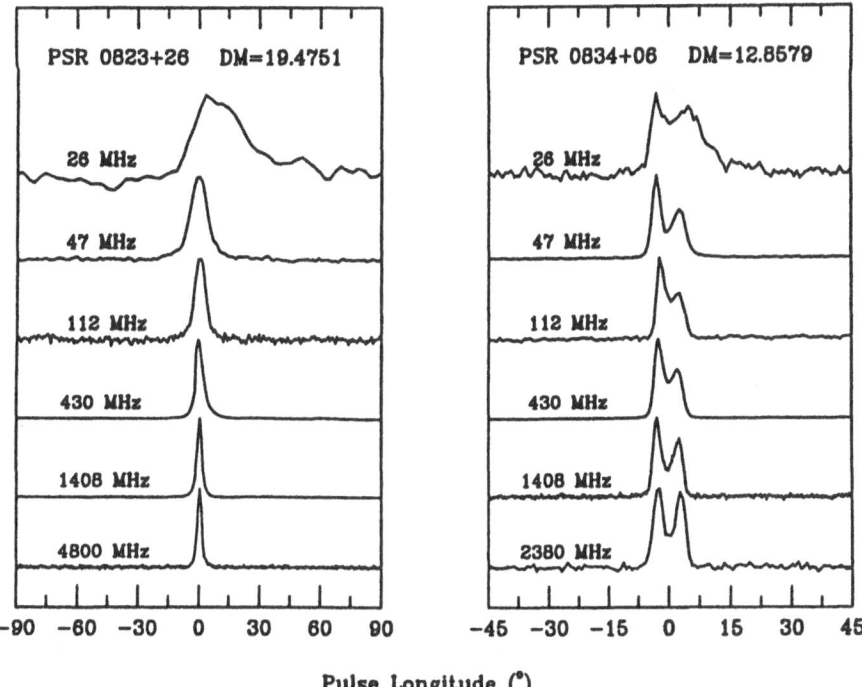

Figure 2: Multi-frequency profiles from Arecibo. The profiles are time-aligned according to the DMs given in each frame. Figure 2 is continued on the next page.

PSR 1604-00, there is substantial profile evolution with frequency – these are discussed in detail by Phillips and Wolszczan (1990).

Topocentric times of arrival (TOAs) were measured by the standard technique of cross-correlating a high signal-to-noise template with the sample profiles from each observing session. The templates we used were the grand average profiles displayed in Figure 2 with the fiducial points shifted to a common pulse longitude. Topocentric TOAs were converted to barycentric arrival times using the MIT solar system ephemeris.

At 25 MHz four of the pulsars display an exponential scattering tail produced by multi-path propagation in the ISM (PSR 0823+26, PSR 0834+06, PSR 0919+06, PSR 1604-00). In those profiles, arrival times of the fiducial points were delayed by interstellar scattering. Using standard techniques (Williamson 1974) we deconvolved an exponential scattering function from each template to find the intrinsic pulse shape. By comparing the "intrinsic" and observed pulse shapes we estimated the scattering delays. The delays were then subtracted from 25 MHz TOAs to correct for the effects of interstellar pulse broadening.

DMs were measured at each frequency with respect to 430 MHz by a least-squares fit of equation (1) to the arrival time data. Figures 3a f show the derived dispersion measures. Except for PSR 0834+06 (Fig. 3b), the DMs appear to be constant over a 200:1 frequency range (50:1 for PSR 1604-00). The dispersion measure for PSR 0834+06

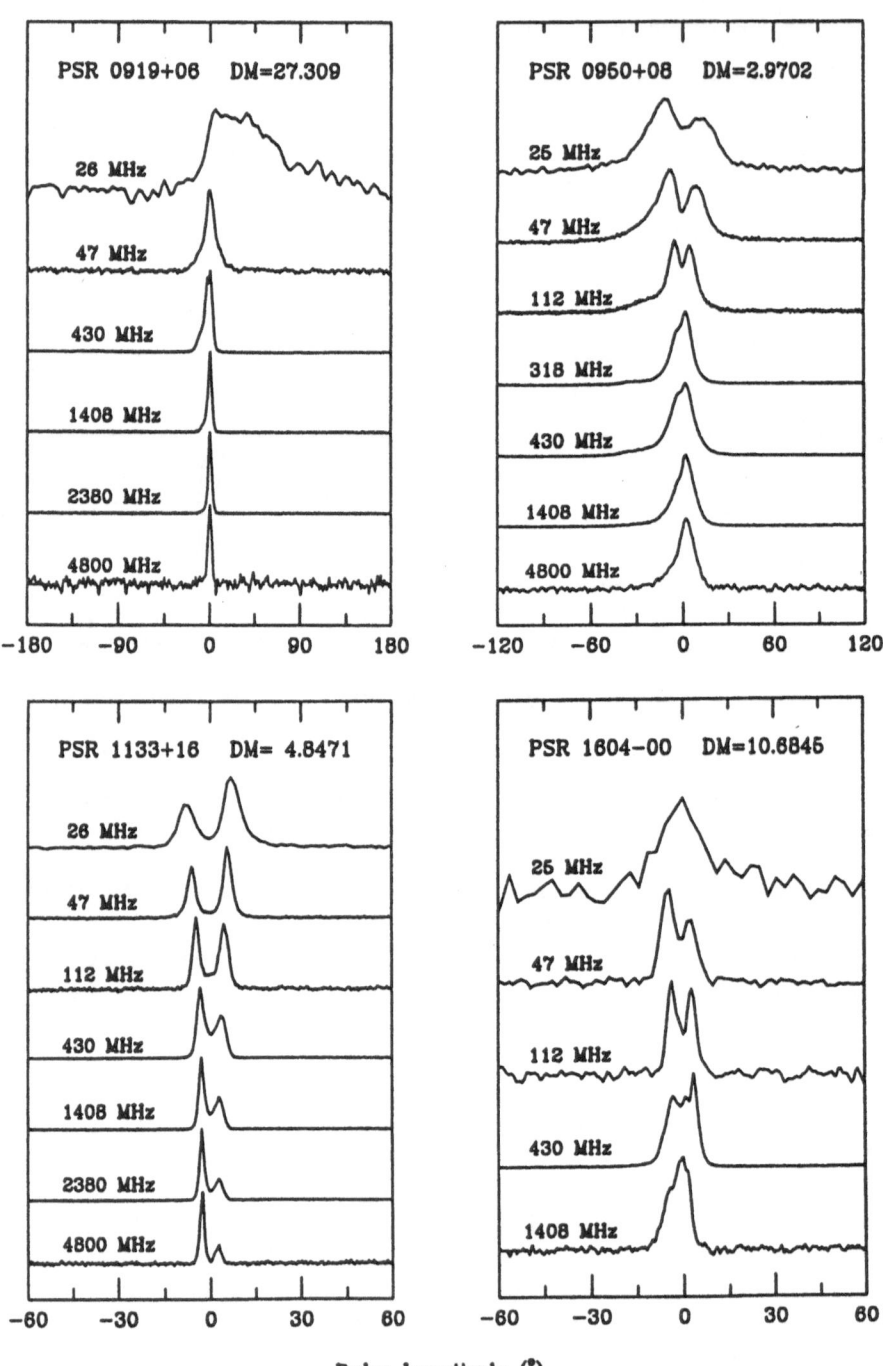

Figure 2 continued: Time-aligned multi-frequency profiles.

is constant from 25 to 1408 MHz but at 2380 MHz it increases significantly. So far we cannot verify this trend at 4800 MHz because the pulsar is too weak to detect. If the *DM* anomaly proves to be real, it could be a result of non-dipolar magnetic field structure near the stellar surface.

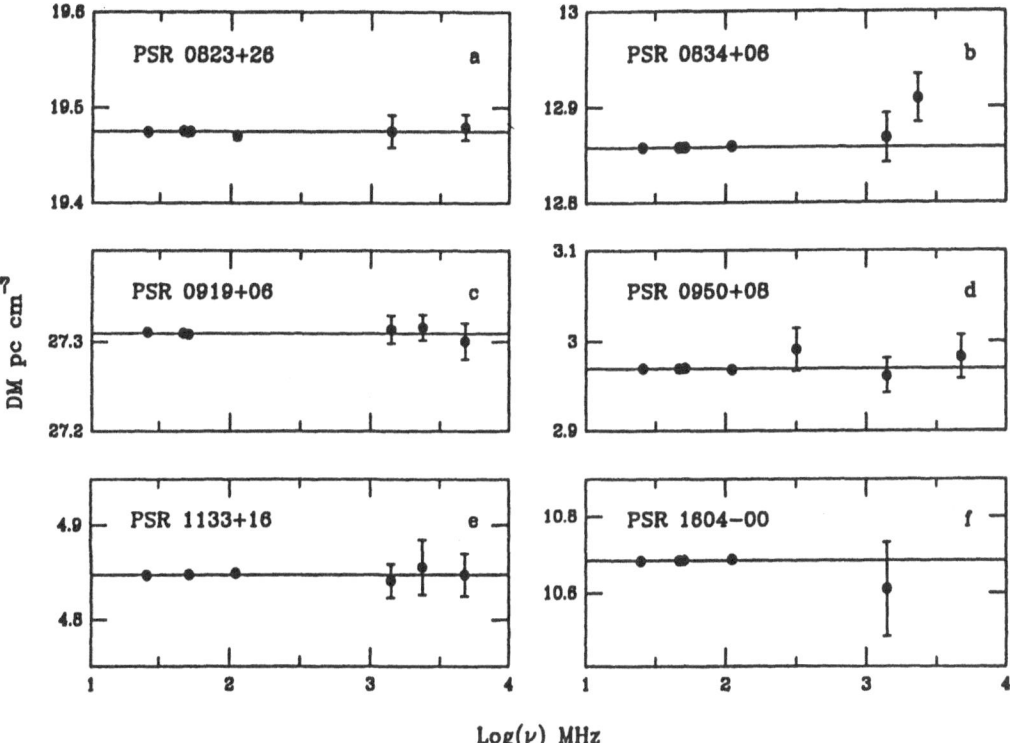

Figure 3: *DM* vs log(ν). *DM* at each frequency was measured with respect to 430 MHz. The horizontal line in each frame represents the weighted mean *DM*. In figure 3b, the highest frequency point – 2380 MHz – was excluded from the average for reasons discussed in the text.

The best dispersion measures were computed from a weighted average of the measurements in Figure 3. Table 1 shows the resultant *DM*s which are the most accurate to date for long period (i.e., non-millisecond) pulsars.

3.3 Superdispersion

The dispersion measure data in Figure 3 clearly show that pulsar arrival times obey a cold plasma dispersion law. Our findings differ form those of Shitov and Malofeev (1985), Shitov et al. (1988) and Izvekova et al. (1989) who report an increase in *DM* at low frequencies for four of the pulsars in our sample (PSR 0823+26, PSR 0834+08, PSR 1133+16, and PSR 1604-00). They attributed the apparent enhancement in *DM* to magnetic field sweepback at high altitudes in the pulsar magnetosphere.

The discordant results lead to very different conclusions regarding the spatial structure of the neutron star's radio source. If the Pushchino data are correct, they imply that the emission region is broadly distributed with high frequencies originating at low magnetospheric altitudes and low frequencies emanating from regions near the light cylinder ($r_{LC} \sim 5 \times 10^4$ km). As we show in §3.4, however, the Arecibo timing results indicate that the emission region is < 200 km in extent and is probably very close to the neutron star's surface. For studies of the pulsar magnetosphere it is clearly of interest to know which scenerio is correct.

Shitov et al. (1988) observed five pulsars at 102.5, 60, 40, and 30 MHz. They aligned their profiles using DMs from the literature and found residual delays ("superdispersion") which increased with decreasing frequency. The simplest explanation for the extra delays is that the dispersion measures they adopted were inaccurate. In four of the five cases, the DMs employed by Shitov et al. were measured 17 years before their own observations (DM for PSR 1133+16 was measured only five years earlier). Multi-frequency observations of millisecond pulsars indicate that DM may vary substantially on ~ 1 yr time scales as ionized interstellar clouds drift by the line of sight (Rawley 1986; Foster et al. 1989; Cordes et al. 1990). To check that simple DM errors could account for superdispersion, we computed the frequency scaling of the delays at 102.5, 60, 40, and 30 MHz given in Table I of Shitov et al. (1988). In each case we found $\tau \propto \nu^{-2}$, as expected for interstellar dispersion. Bruck et al. (1986) have timed PSR 0809+74, the pulsar with the largest reported superdispersion, and find that it also obeys a cold plasma dispersion law between 17 and 25 MHz. The same result was obtained by Davies et al. (1984) using multi-frequency data between 406 and 39 MHz (see also Popov 1987). Based on the evidence discussed in this and the previous section, we conclude that no deviations from cold plasma dispersion have been detected in low-frequency TOA measurements.

3.4 Emission radii

The method of using multi-frequency timing to constrain emission radii was first discussed by Cordes (1978). We have extended his analysis to include the effects of magnetic field sweepback and, using our data, derived strong constraints on the size of the pulsar emission region.

If frequencies ν_1 and ν_2 are emitted from radii r_1 and r_2 in the pulsar magnetosphere, there will be an arrival time delay $t(\nu_2) - t(\nu_1)$ consisting of the following three terms:

Retardation. The path length difference $r_2 - r_1$ introduces a delay

$$t(\nu_2) - t(\nu_1) = \frac{P}{2\pi} \left(\frac{r_1 - r_2}{r_{LC}} \right) \equiv \Delta t_r \tag{2}$$

where $r_{LC} = cP/2\pi$ is the light cylinder radius.

Abberation. Because the pulsar is rotating, the radiation beam at altitude r will be bent by an angle

$$\theta_a = \tan^{-1} \left(\frac{v_\phi}{c} \right) = \tan^{-1} \left(\frac{r \sin \alpha}{r_{LC}} \right)$$

where $v_\phi = r\Omega$ is the azimuthal corotation velocity and α is the angle between the rotational and magnetic-dipole axes. The beam deflection is in the direction of the pulsar rotation and increases with altitude. The time delay from differential beam bending ("abberation") is

$$t(\nu_2) - t(\nu_1) = \frac{P}{2\pi} \left[\tan^{-1}\left(\frac{r_1 \sin\alpha}{r_{LC}}\right) - \tan^{-1}\left(\frac{r_2 \sin\alpha}{r_{LC}}\right) \right] \equiv \Delta t_a \qquad (3)$$

Magnetic sweepback. Pulsars lose rotational energy in the form of magnetic dipole radiation. The electromagnetic torque on the neutron star results in a toroidal bending of the magnetic field opposite to the sense of the star's rotation. At radius r the bending angle is $\theta_{mfs} \simeq 1.2(r/r_{LC})^3 \sin^2\alpha$ for a spinning magnetic dipole (Shitov 1983). Assuming that radiation beams are aligned with the local magnetic field, differential beam bending gives a time delay

$$t(\nu_2) - t(\nu_1) = 1.2 \sin^2\alpha \frac{P}{2\pi} \left[\left(\frac{r_2}{r_{LC}}\right)^3 - \left(\frac{r_1}{r_{LC}}\right)^3 \right] \equiv \Delta t_{mfs} \qquad (4)$$

From equations (2) through (4) we see that retardation/abberation and magnetic field sweepback (MFS) act in opposite ways. Retardation and abberation delay the low altitude frequencies while MFS delays the high altitude frequencies.

After correcting for interstellar dispersion, the total delay is

$$\Delta t'(r(\nu_1), r(\nu_2)) = \Delta t_r + \Delta t_a + \Delta t_{mfs} \qquad (5)$$

To simplify the analysis we have ignored the possibility of additional terms due to multipole magnetic field structure and interstellar refraction. The accurate correspondence of TOAs to a cold-plasma dispersion law means that $\Delta t'$ is undetectable at the current level of precision. However, the timing data can be used to set interesting limits on $\Delta r = r_2 - r_1$ by solving the equation

$$\Delta t'(r_1, r_2) \leq \epsilon_{max}(\nu_1, \nu_2), \qquad (6)$$

where ϵ_{max} is the maximum delay between ν_1 and ν_2 after correcting for interstellar dispersion:

$$\epsilon_{max} = \left[t(\nu_2) - t(\nu_1) - 4.148 \times 10^9 DM(\nu_2^{-2} - \nu_1^{-2}) \right]_{max}$$

Values for ϵ_{max} are listed in Table 1. We have chosen 47 MHz as the low frequency for this analysis because the time resolution was 2 to 5 times better than at 25 MHz.

Compared to the light travel time across the magnetosphere ($r_{LC}/c \sim 150$ ms), ϵ_{max} (~ 1 ms) is small . The possible explanations for the relatively low limit on $\Delta t'$ are: (1) $r_2 - r_1$ may be large but retardation/abberation and MFS delays cancel, or (2) $r_2 \approx r_1$, meaning that each term in equation (5) is independently small. We can rule out the first alternative because simultaneous cancellation between all the frequency pairs we observed would require a highly artificial radius-to-frequency mapping. The second alternative is consistent with our data and is the more plausible.

The delay $\Delta t'$ is a 2-dimensional function of r_1 and r_2 or, equivalently, r_1 and Δr. We numerically inverted equation (6) to find $\Delta r(r_1)$ for each pulsar. Representative

Table 1: Pulsar DMs and constraints on emission radii

PSR	DM	ν_1	ν_2	ϵ_{max}	Δr_{max}	$r_{1,max}$
	pc cm^{-3}	MHz	MHz	μs	km	km
0823+26	19.4751 $\pm$ 0.0002	4800	47	1400	210	20
0834+06	12.8579 $\pm$ 0.0002	1408	47	1100	200	*
0919+06	27.309 $\pm$ 0.001	4800	47	1100	200	560
0950+08	2.9702 $\pm$ 0.0001	4800	47	700	210	30
1133+16	4.8471 $\pm$ 0.0002	4800	52	900	150	40
1604-00	10.6845 $\pm$ 0.0003	430	47	900	150	570

results are displayed in Figure 4. In general, Δr_{max} increases from a minimum near the stellar surface to a maximum at higher altitudes where magnetic sweepback cancels retardation and abberation. The maxima are double-peaked because solutions to the equation $\Delta t_a(r_1, r_2) + \Delta t_r(r_1, r_2) = -\Delta t_{mfs}(r_1, r_2)$ are symmetrical in r_1 and r_2. As we discussed above, solutions based on the mutual cancellation of terms in equation (5) can be discarded. If we ignore the MFS term in equation (5), thereby removing the possibility of cancellation, we find $\Delta r \leq 200$ km for the pulsars in our sample (see Table 1). This is identical to the limit on Δr one would derive at low altitudes if all the terms in (5) were kept (cf., Figure 4).

The timing data constrain Δr but *not* r_1. The altitude can be estimated if we assume: (1) that the magnetic field is dipolar and (2) that the pulse width at frequency ν is determined by the angular width of the radiation zone at $r(\nu)$. The "radiation zone" is defined to be the flux tube of field lines which pierce the light cylinder and thread the magnetic polar cap of the neutron star. From Cordes (1978),

$$r_1 \leq \frac{\Delta r_{max}}{(\theta_2/\theta_1)^2 - 1} \qquad (7)$$

where θ_2 (θ_1) is the pulse width at the lowest (highest) frequency. Table 1 shows the maximum radii calculated using equation (7). (PSR 0834+06 was excluded from the analysis due to its unusual profile evolution: $\theta(\nu)$ is minimum at 111.5 MHz and increases towards both lower and higher frequencies. Normally $\theta(\nu)$ increases monotonically from high to low frequencies.) The best limits indicate that the 5 GHz emissions come from a region less than 1 to 3 stellar radii above the neutron star surface.

Our results show that the pulsar emission region is radially compact and is probably situated close to the stellar surface. This picture is supported by multi-frequency timing of the millisecond pulsar PSR 1937+21 (Cordes and Stinebring 1984). That object has a light cylinder radius of only 78 km, so all the observed emission must come from within $\sim$ 70 km of the neutron star's surface. Furthermore, a retardation analysis of TOAs at three frequencies (.32, .43 and 1.4 GHz) showed that Δr was less than 2 km. That limit is $\sim$ 2 orders of magnitude smaller than the ones listed in Table 1.

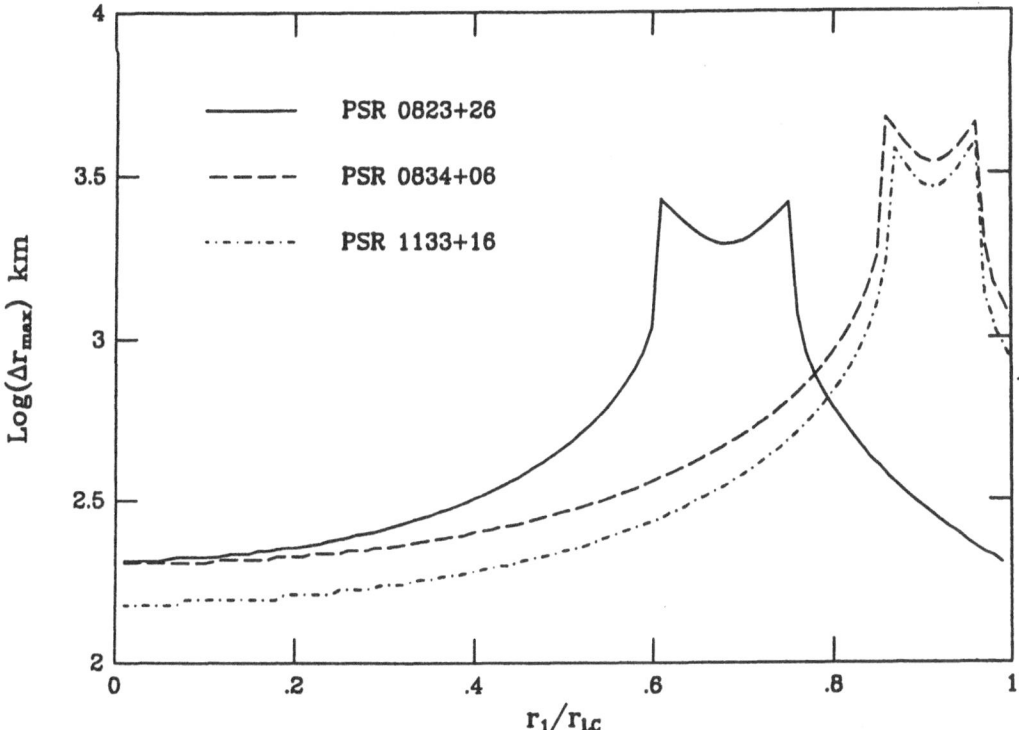

Figure 4: Representative plots of Δr_{max} vs. emission radius r_1 for three pulsars.

If emission mechanisms for millisecond and "normal" pulsars are the same, it seems that all frequencies may come from the same altitude, at least over the 4:1 frequency range reported by Cordes and Stinebring. Additional timing observations of millisecond pulsars covering a broader range of frequencies would be useful in probing the radius-to-frequency mapping in pulsar magnetospheres. As for long-period pulsars, the limits in Table 1 and Figure 4 may represent the best possible using the multi-frequency timing method.

This work was supported by the National Astronomy and Ionosphere Center at Cornell University which operates the Arecibo Observatory under contract with the National Science Foundation.

References

Bruck, Yu.M. 1987, *Australian J. Phys.*, **40**, 861.

Bruck, Yu.M. 1989, private communication.

Bruck, Yu.M. and Ustimenko B.Yu. 1976, *Nature*, **260**, 766.

Bruck, Yu.M. and Ustimenko B.Yu. 1977, *Astrophys. Space Sci.*, **51**, 225.

Bruck, Yu.M. and Ustimenko B.Yu. 1979, *Astron. Astrophys.*, **80**, 170.

Bruck, Yu.M., Ustimenko, B.Yu., Popov, M.V., Soglasnov, V.A. and Novikov, A.Yu. 1986, *Sov. Astron. Lett.*, **12**, 381.

Cordes, J.M. 1978, *Ap. J.*, **222**, 1006.

Cordes, J.M. and Stinebring D.R. 1984, *Ap. J. Letters*, **277**, L53.

Cordes, J.M., Weisberg, J.M., and Boriakoff, V. 1985, *Ap. J.*, **288**, 221.

Cordes, J.M., Wolszczan, A., Dewey, R.J., Blaskiewicz, M. and Stinebring, D.R. 1990, *Ap. J.*, **349**, 245.

Craft, H.D. 1970, Ph.D. thesis, Cornell University, Ithaca, New York.

Davies, J.G., Lyne, A.G., Smith, F.G., Izvekova, V.A., Kuz'min, A.D. and Shitov, Yu. P. 1984, *Mon. Not. R. Astr. Soc.*, **211**, 57.

Foster, R.S., Backer, D.C. and Wolszczan, A. 1989 (in press).

Izvekova, V.A., Kuz'min A.D., Malofeev, V. M. and Shitov, Yu. P. 1989, in *Pulsars, Proceedings of the Lebedev Physics Institute*, Ed. Kuz'min, A.D., USSR Academy of Sciences, Moscow.

Kuz'min, A.D. 1986, *Sov. Astron. Lett.*, **12**, 325.

Kuz'min, A.D., Izvekova, V.A., Malofeev, V.M., and Shitov Yu. P. 1988, *Sov. Astron. Lett.*, **14**, 58.

Narayan, R. and Vivekanand, M. 1983, *Ap. J.*, **274**, 771.

Phillips, J.A. and Wolszczan, A. 1989, *Ap. J. Letters*, **344**, L69.

Phillips, J.A. and Wolszczan, A. 1990, in preparation.

Popov, M.V. 1987, *Sov. Astron. Letters*, **13**, 41.

Rawley, L. 1986, Ph.D. dissertation, Princeton University.

Reyes, F.R. 1989, Ph.D. dissertation, University of Florida, Gainesville.

Rickett, B.J. 1988, in *Radio Wave Scattering in the Interstellar Medium*, Eds. Cordes, J.M., Rickett, B.J. and Backer, D.C., American Institute of Physics, New York.

Shitov, Yu.P. 1983, *Sov. Astron.*, **27**, 314.

Shitov, Yu.P. and Malofeev, V.M. 1985, *Sov. Astron. Lett.*, **11**, 39.

Shitov, Yu.P., Malofeev, V.M., and Izvekova, V.A. 1988, *Sov. Astron. Lett.*, **14**, 181.

Tanenbaum, B.S., Zeissig, G.A., and Drake, F.D. 1968, *Science*, **160**, 760.

Williamson, I.P. 1974, *Mon. Not. R. Astr. Soc.*, **166**, 499.

Refractive Interstellar Scintillation Effects at Low Frequencies

Yashwant Gupta

University of California, San Diego

Pulsar observations : Although refractive scintillations have been predicted for some time,[1] there are not many conclusive observations from which one can estimate the time scale and modulation index.[2,3] Since January 1989, we have been monitoring pulsar amplitudes at UC San Diego. Nine pulsars have been observed daily for the past 11 months with our 74 MHz antenna at Fallbrook. The antenna is a dipole phased array consisting of 128 1λ by 4λ elements arranged in 32 rows (north-south) and 4 columns (east-west) giving a total extent of 32λ by 16λ. This gives a HPBW of $1.75°$ (n-s) and $3.5°$ (e-w). The effective collecting area is $8000\,m^2$ and the system noise temperature is about $1000°$ K for most regions of the sky. The IF bandwidth used is 500 kHz, which is large enough to quench the diffractive scintillations since the typical decorrelation bandwidth at 74 MHz is only a few hundred hertz. After detection, the data are low pass filtered with a cut off of 20 Hz and digitally sampled at 50 Hz. The pulsars are observed once every day for periods ranging from 20 to 40 minutes. Each day's data is coherently averaged using the doppler corrected period of the pulsar to obtain a daily pulse profile. An estimate for the pulsar strength is obtained by comparing this daily pulse profile with a standard template for the pulsar. This estimate is corrected for gain fluctuations of the antenna system using observations of several strong continuum sources. The final result is a intensity time series for each pulsar with samples once every day. These are displayed in Figs. 1a and 1b.

Most of the pulsars observed are nearby, since we expect these to show the shortest time scales of refractive fluctuations. For comparison, some relatively further ones are also observed. The expected refractive time scale, in MKS units, is given by [1,4]

$$\tau_r = 2.3 \times 10^{-18} \, \lambda^{2.2} \, z^{1.6} \, C_n^{1.2} \, V_r^{-1} \tag{1}$$

where the parameter C_n^2 is estimated using

$$C_n^2 = 6.5 \times 10^{35} \, z^{-1.83} \, \lambda_{obs}^{-3.67} \, \delta v_d^{-0.83} \tag{2}$$

These calculations are based on the thin screen model for interstellar scattering and assume a Kolmogorov type power law spectrum with a slope of 11/3 for the electron density fluctuations in the ISM. The values for the decorrelation bandwidth, δv_d, at frequency, v_{obs}, are taken from literature.[4] Values for the relative velocity, V_r, are taken from proper motion studies.[5] Z is the distance to the pulsar and is taken from a standard table of pulsar parameters.[6] The calculated time scales range from 30 days to 450 days. These predicted values are shown in Table I. It should be noted that the uncertainty in the values of V_r is a major cause for errors in the theoretical estimates of the refractive time scales.

As can be seen from the data of Fig. 1, the pulsars exhibit a wide range of time scales of fluctuations. For some pulsars, like P1133+16, the observed fluctuations appear to have a time scale (˜ 30 days) which matches fairly well with the expected value. For pulsars like P0834+06 and P1508+55, though there are not enough refractive "scintles", the apparent time scale does seem to match the expected value. Also, pulsars which have large expected time scales (like P0329+54 and P1919+21) remain relatively steady over the span of the current data set, which is also consistent with our calculations. For pulsars like P0950+08, data and calculations appear to be in disagreement. However, the proper motion velocity for this pulsar is very small and corrections to the relative velocity due to motion of the observer need to be investigated. For most of the pulsars, substantially longer spans of data are needed to make quantitative estimates of the time scales.

Low frequency effects : Using the theoretical model, we can predict the frequency behaviour of the time scale and modulation index of the refractive process. The relevant expression for the time scale is given in equation (1) and that for the modulation index is given by [1]

$$m_r^2 = 1.4 \times 10^{11} \; \lambda^{-1.13} \; z^{0.067} \; C_n^{-0.8} \tag{3}$$

Thus at lower frequencies, we should expect to see reduced refractive fluctuations and longer time scales with the effect being more prominent for the time scale (see table II). Hence for pulsars and other point like sources, the time scale should become very long (˜50 years), so that the refractive fluctuations will become unimportant at low frequencies.

The above results are valid only for point sources. For extended sources, the ratio of the source diameter (θ_s) to the scattering angle (θ_d) affects both the modulation index (m_r^e) and time scale (τ_r^e). These are given as [7]

$$m_r^e = m_r \theta_d \; / \; (\theta_d^2 + \theta_s^2)^{0.5} \tag{4}$$

$$\tau_r^e = \tau_r (\theta_d^2 + \theta_s^2)^{0.5} \; / \; \theta_d \tag{5}$$

where m_r and τ_r are the point source values. These results apply for extragalactic sources at high galactic latitude and ignore the enhanced scattering at lower galactic latitudes. For typical values of z=500 pc, $C_n^2 = 10^{-3.5}$, and a fixed source size of 10 mas, the results for different frequencies are displayed in table II. When θ_s dominates θ_d, the effective modulation index is decreased and the time scale is increased.

For most sources, θ_s is a function of frequency. A linear increase of source size with wavelength is a reasonable assumption for many sources.[7] Using this modification yields the results shown in table III. These predict even smaller modulation indices and even larger time scales at lower frequencies.

Thus we conclude that refractive scintillation from the ISM will not play a significant role in observations at low frequencies. However, the diffractive scatter broadened diameter, θ_d, for most sources will be very large ($\sim$ arcmin). As has been emphasised by earlier speakers in this meeting, this could restrict angular resolution of observations made at low frequencies.

The author would like to thank Dr.B.J.Rickett for valuable discussions and advice and NRL for partial travel funding. Some of the work was supported by a grant from the California Space Institute.

References :

1. Rickett, B.J., Coles, W.A., Bourgois, G. 1984. Astron. Astrophys. **134** : 390-395
2. Cole, T.W., Hesse, H.K., Page, C.G. 1970. Nature **225** : 712-713
3. Huguenin, G.R., Taylor J.H., Helfand, D.J. 1973. Ap.J. **181** : L139-L142
4. Coles, W.A., Frehlich, R.G., Rickett, B.J., Codona, J.L. 1987. Ap.J. **315** : 666-674
5. Lyne, A.G., Anderson B., Salter M.J. 1982. MNRAS. **201** : 503-520
6. Manchester, R.N., Taylor, J.H. 1981. Astron. J. **86** : 1953-1973
7. Rickett, B.J. 1986. Ap.J. **307** : 564-574

Table I

Calculation of predicted refractive time scales

Pulsar	Z (kpc)	$\delta\nu_d$ (kHz)	ν_{obs} (MHz)	C_n^2 ($m^{6.67}$)	V_r (km s^{-1})	τ_r (days)
P0329+54	2.30	82	410	6.6×10^{-5}	229	446
P0809+74	0.17	121	156	1.6×10^{-4}	41	67
P0823+26	0.71	360	430	2.0×10^{-4}	365	82
P0834+06	0.43	260	430	6.5×10^{-4}	104	266
P0950+08	0.09	1000	320	1.3×10^{-3}	15	223
P1133+16	0.15	800	430	1.8×10^{-3}	264	35
P1237+25	0.33	729	430	4.5×10^{-4}	178	81
P1508+55	0.73	160	340	1.6×10^{-4}	346	79
P1919+21	0.33	160	430	1.6×10^{-3}	100	307

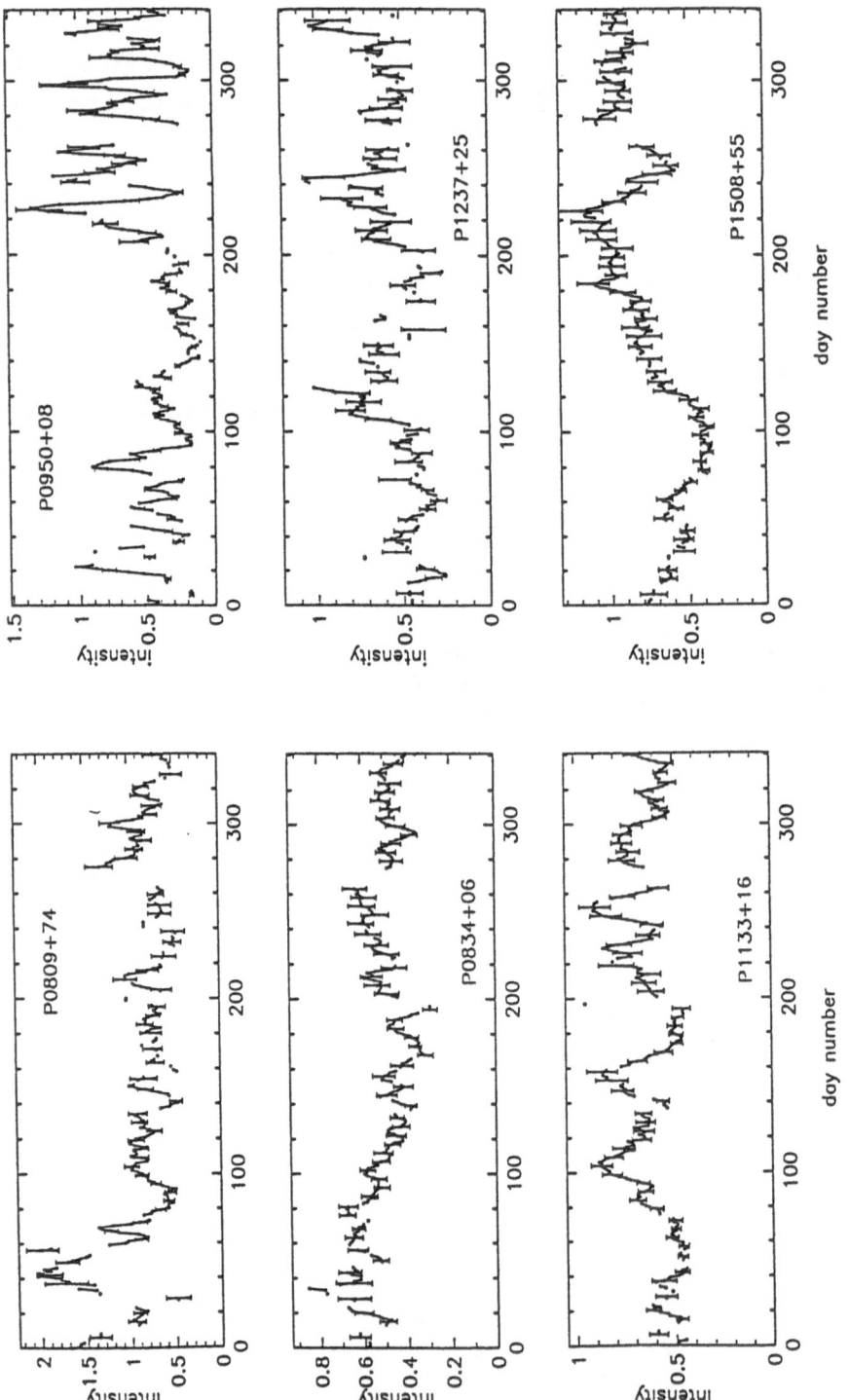

Fig. 1a : Long term intensity fluctuations for six pulsars at 74 MHZ.
Intensity values are normalised to arbitrary units. The data has been
smoothed with a 5 day running mean.

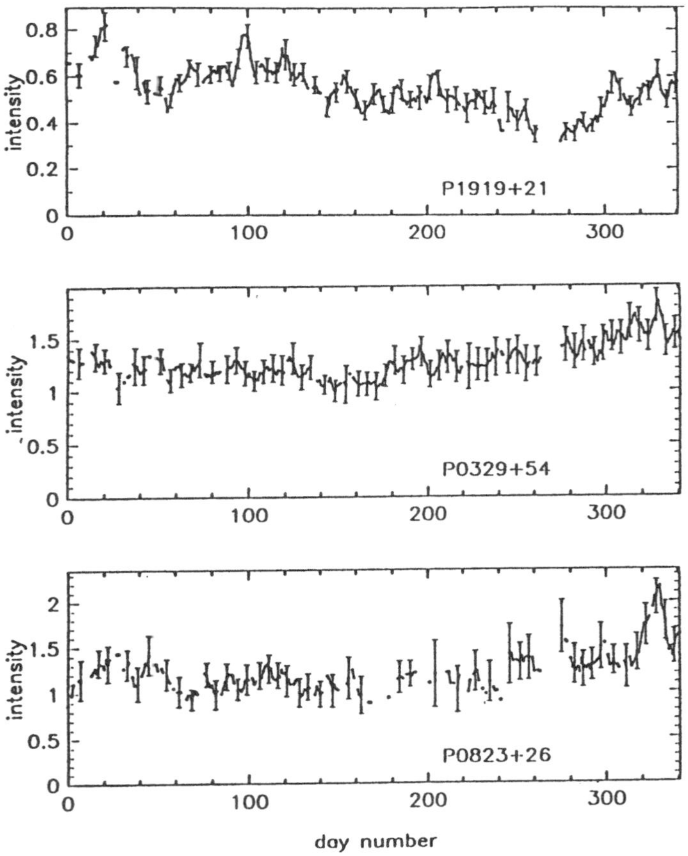

Table II : Effect of finite source size

Freq (MHz)	θ_s (mas)	θ_d (mas)	m_r^p	m_r^e	τ_r^p (days)	τ_r^e (days)
3000	10	1.5×10^{-2}	0.68	0.001	1.3×10^{-1}	8.7×10^1
300	10	2.4×10^1	0.18	0.042	2.1×10^1	8.9×10^1
30	10	3.8×10^2	0.05	0.05	3.3×10^3	3.3×10^3
3	10	6.0×10^4	0.013	0.013	5.2×10^5	5.2×10^5

Table III : Effect of frequency dependent finite source size

Freq (MHz)	θ_s (mas)	θ_d (mas)	m_r^p	m_r^e	τ_r^p (days)	τ_r^e (days)
3000	10	1.5×10^{-2}	0.68	0.001	1.0×10^{-1}	8.7×10^1
300	100	2.4×10^1	0.18	0.024	2.1×10^1	8.7×10^2
30	1000	3.8×10^2	0.05	0.018	3.3×10^3	9.3×10^3
3	10000	6.0×10^4	0.013	0.013	5.2×10^5	5.3×10^5

Fig. 1b : : Long term intensity fluctuations for three pulsars at 74 MHz. Intensity values are normalised to arbitrary units. The data has been smoothed with a 5 day running mean.

VI. SURVEYING AT LOW FREQUENCIES

LOW FREQUENCY SURVEYING

J.E. Baldwin

Cavendish Laboratory, Madingley Rd., Cambridge CB3 0HE

1. Introduction

The essence of surveying is to cover a lot of ground and to find out where things are. At high frequencies in radio astronomy, say above 1 GHz, surveying is an activity usually completely separate from that of high resolution imaging, requiring a different assessment of what is scientifically interesting and a different type of telescope. But at low frequencies there is no clear separation of functions; the maximum resolution achievable is poorer and the field of view is inevitably large. So any low frequency telescope for high resolution imaging is also a survey telescope. The history of low frequency arrays in the 1960s, 70s and 80s illustrates this; all of the low frequency telescopes such as the Cambridge 38 MHz array (Williams et al 1966), the DRAO 10 MHz (Bridle & Purton, 1968) and 22 MHz (Roger et al, 1969) arrays, the Culgoora array at 80 and 160 MHz (Slee and Higgins 1973; Slee 1977) the Clark Lake array (Viner and Erickson, 1975) the Cambridge 151 MHz 6C (Baldwin et al, 1985) and 38 MHz arrays (Rees, 1990) provided the highest resolutions and sensitivities achieved at those frequencies and at the same time were survey instruments. We can expect the same to be true to an even greater extent for very low frequency arrays.

Assuming that it is so, there are three main questions to answer. First, what are the ultimate limits on the sensitivity and resolution of low frequency telescopes; second, given these constraints, what interesting science is possible and thirdly, how might the ultimate arrays be designed and constructed. Of particular interest here are the relative merits of ground-based and space-based arrays.

2. Constraints on resolution and sensitivity

The main limitations on low frequency observations have already been mentioned in this workshop. They are interplanetary and interstellar scattering, radio interference and money.

The worst effects of interplanetary scattering can be avoided by looking in the antisolar direction. Interstellar scattering then dominates and gives broadening of point sources into scattered disks whose diameters are ~0.16 arcsec at 81.5 MHz increasing with wavelength a little faster than λ^2. In the range of frequencies of interest here the sizes are roughly:

Frequency MHz	Resolution	Max. useful baseline km
30	1 arcsec	2000
10	10 arcsec	600
3	2 arcmin	160
1	20 arcmin	50

An immediate remark on this table is that it looks difficult to design a telescope which matches the requirements over the whole range of frequencies. The baselines required are such that arrays could be built on the Earth, in space or even on the Moon.

The second important constraint is the narrow bandwidth imposed on any space array (and many ground-based arrays) by the radio emission spectrum of the Earth. There are a vast number of powerful broadcast stations operating at both legal and illegal frequencies plus a similarly vast number of less powerful unlisted fixed and mobile transmitters. Lists of broadcast stations in the mid-1980s showed that below 12 MHz there were only four gaps as wide as 100 kHz between centre frequencies of stations. In some parts of the spectrum the total emission exceeds 250 MW in a 100 kHz band. This corresponds to a flux density of the Earth seen from a distance of 100,000 km of $>3 \times 10^{12}$ Jy in that band! More worrying is that even a mean level of 1 W Hz^{-1} radiated from the Earth gives 10^9 Jy at that distance. Since there is a requirement to reach flux densities of <0.1 Jy at 30 MHz and ~1 Jy at 3 MHz, there is clearly a serious problem; no array built so far gives a dynamic range even as large as 10^6. It is hard to escape the conclusion that a space array is not feasible at any frequency in the range 1-30 MHz for which the ionosphere is transparent. In practice that applies at all times for frequencies >10 MHz.

The question of where, and how, satisfactory arrays can be built is taken up in Section 5.

3.   Radio source surveys at low frequencies

A central question in assessing the value of low frequency source surveys is whether or not we expect to see many or any new sources. A good starting point for this discussion is Rees 38 MHz survey. What fraction of these sources will appear in a deep survey at, for example, 3.8 MHz? Fig 1 shows the distribution of spectral indices between 38 and 151 MHz, for sources with S>1 Jy selected at 38 MHz, derived by Mark Lacy. The distribution is very close to a Gaussian in shape with a mean spectral index $<\alpha>$ of 0.788 and a standard deviation of 0.196. Only a few points lie at extreme values of α. If the spectra of all sources are straight and the distribution of α is Gaussian with standard deviation σ at some frequency ν_0, then an old but useful algebraic result (Long 1963, Williams & Bridle 1967, Kellermann 1964) tells us that for a complete sample chosen at some other frequency ν, the distribution of spectral indices is also Gaussian with the same standard deviation and a mean value

$$\langle\alpha(\nu)\rangle = \langle\alpha(\nu_0)\rangle + \mu\sigma^2\ell n(\nu/\nu_0)$$

where μ is the slope of the integral source counts at the limiting flux density.

So we may expect that

$$\langle\alpha(3.8)\rangle = 0.788 + (-1.6)\ (0.196)^2\ \ell n(3.8/38)$$
$$= 0.935$$

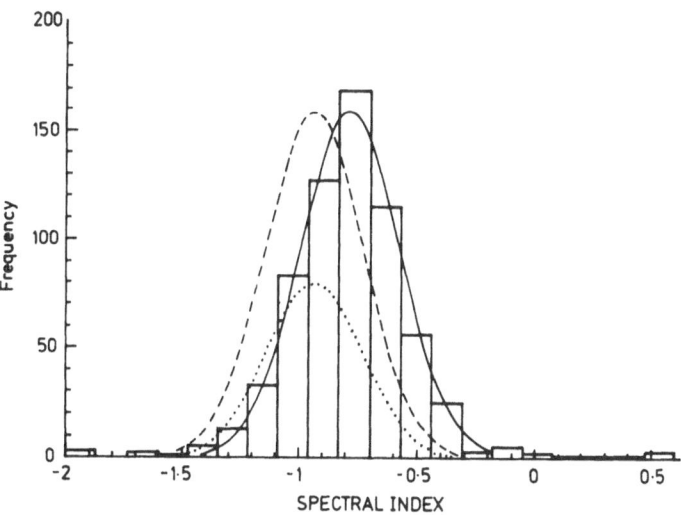

Fig 1   Histogram of spectral indices of sources selected at 38 MHz.
Thin full line   :   gaussian fit to the observed histogram
Dashed line   :   expected spectral index distribution at 3.8 MHZ for a
                                      sample of equal depth
Dotted line   :   for a sample with half the number of sources

The dashed curve in Fig 1 shows the expected distribution of α for a sample of the same size as at 38 MHz. From the overlap in the curves it is clear which sources are common to the samples and which are 'new' at 3.8 MHz. Not suprisingly they are the weaker sources with spectral indices steeper than the mean. If, however, the depth of the survey at 3.8 MHz is poorer, say by a factor of 2 in numbers, then the distribution is as shown in the dotted Gaussian curve with half the total number of sources. Note that in this case essentially none of the sources at 3.8 MHz are new sources. This is a rather obvious point to labour but it emphasizes that, if we are to expect new things at very low frequencies, then the sensitivity of telescopes is as vital a factor as it is at higher frequencies. Put another way, it says that the range of spectral indices which we have in existing surveys is not very large; there are very few sources with dramatically steep spectra which otherwise might dominate samples at very low frequencies. That is not to say that there are none or that they are unimportant. Potentially they are very exciting; we already know of msec pulsars with very steep spectra; other sources of coherent emission are imaginable which might well justify the whole programme even if they are few in number.

All this discussion presupposes that source spectra are straight. Whilst this is approximately true for many sources, there are strong reasons to expect large departures at low frequencies. In particular, the effects of thermal absorption in both the ISM and in regions close to the radio sources and of synchrotron self-absorption should be important. Note that both of these effects will reduce the numbers in low frequency samples below those discussed above. Other speakers will be discussing thermal absorption; I will deal only with synchrotron self-absorption.

4. Synchrotron Self-Absorption in Radiosources at Low Frequencies

The importance of synchrotron self-absorption at low frequencies is that it is likely to affect virtually all extragalactic sources, not just those very compact sources which commonly show the effect at giga-Herz frequencies. An analysis by Simon (1977) concluded that the spectrum of the extragalactic background radiation, peaking near 3 MHz, could be accounted for by the integrated effect of the population of known radio sources. She made use of the known structures of sources at high frequencies to calculate the frequency at which self-absorption would occur for each source and summed their effects. The question I raise here is whether the known structure of a source *does* enable one to predict precisely the form of the low frequency spectrum. If it does, then the low frequency measurements would merely confirm what we expect. But if not, then we may get new information that cannot be obtained in any other way. It turns out interestingly that the frequency of turn-over in the spectrum is sensitive only to the filling factor of the radiation in the source. This follows from some of the basic formulae from synchrotron radiation theory.

Pacholczyk (1970, p. 171) gives an expression for the value of the magnetic field H_{min} for which the energy in a source is a minimum for given source parameters

$$H_{min} = (4.5)^{2/7} (1+k)^{2/7} C_{12}^{2/7} \varphi^{-2/7} R^{-6/7} L^{2/7}$$

where R is the radius of a supposed spherical source,
L is the source luminosity,

$$L = 4\pi D^2 \int_{v_1}^{} S(v)dv$$

(1+k) is the factor by which the total energy of the relativistic electrons and protons exceeds that in the electrons alone or, alternatively, is a measure of the departure from the minimum energy condition and φ is the filling factor of the source. Replacing the spherical source at distance D by a rectangular box of transverse cross section ΩD^2 and depth ℓ in the line of sight, and substituting

$$\frac{4}{3} \pi R^3 = V = \Omega D^2 \ell$$

$$H_{min} = \left\{ 4.5 \ (1+k) \ C_{12} \ \varphi^{-1} \ \frac{16\pi^2}{3} \ \frac{S(v_0)}{\Omega\ell} \cdot v_0^\alpha \ \frac{[v_2^{1-\alpha} - v_1^{1-\alpha}]}{(1-\alpha)} \right\}^{2/7}$$

where the radio spectral index between frequencies v_1 and v_2 is α. Note that here we have tried to express H_{min} as a function of directly measureable variables as far as possible. Only ℓ, the thickness of the source in the line of sight, $(1+k)$ and φ are significant unknowns. The second important expression is that for the absorption coefficient for synchrotron radiation, k_v, (Pacholczyk, p. 96)

$$k_v = C_6(\gamma) \ N_0 \ (H\sin\theta)^{(\gamma+2)/2} \ (v/2c_1)^{-(\gamma+4)/2} \ ,$$

where the energy spectrum of the electrons is given by

$$N(E) = N_0 E^{-\gamma} \ .$$

Substituting for N_0 in the expression for k_v as a function of H_{min} through the minimum energy equation leads to

$$k_v = A(\gamma) \ (1+k)^{(2\gamma-3)} \ \varphi^{-(4+2\gamma)/7} \ (S(v_0)v_0{}^\alpha/\Omega\ell)^{(4+2\gamma)/7} \ v^{-(\gamma+4)/2}$$

or, more conveniently, the frequency at which the optical depth through the source is unity is

$$v_{\tau=1} = \{A(\gamma) \ (1+k)^{(2\gamma-3)/7} \ \varphi^{-(4+2\gamma)/7} \ (S(v_0)v_0{}^\alpha/\Omega)^{(4+2\gamma)/7} \ \ell^{(3-2\gamma)/7}\}^{2/(\gamma+4)} \ .$$

The interesting result is that for γ lying in the interval $2 \leqslant \gamma \leqslant 2.8$, corresponding to the common range of spectral indices $0.5 \leqslant \alpha \leqslant 0.9$, the dependence on both $(1+k)$ and ℓ is very weak; $0.0476 \leqslant (4\gamma-6)/7(\gamma+4) \leqslant 0.1092$. An uncertainty in either of these quantities of even a factor of 100 leads to a change in the frequency for optical depth unity of only 1.24 - 1.65. However the power dependence on the filling factor φ is much stronger; $0.3810 \leqslant 4(2+\gamma)/7(\gamma+4) \leqslant 0.4034$. This is of some importance since sources often show filamentary structure down to the limits of resolution (eg Cyg A) and we do not know at present how far that structure extends to even finer scales.

The effectiveness of the method can of course be tested at high frequencies for sources whose structure is well determined and which show self-absorption effects in their spectra. An example which looks suitable for analysis is 3C295. Its overall spectrum (Fig 2) shows a peak just below 100 MHz. It lies at high galactic latitude and is very unlikely to show any effects of thermal absorption. It has been well mapped at a number of frequencies.

Fig 3 (Laing, personal communication) shows the 4.9 GHz structure with a resolution of 0.3 arcsec. The peak surface brightness is 1.35 Jy/beam at this frequency and resolution. If the depth in the line of sight is assumed to be the same as the lateral scale and $(1+k)$ and φ are both unity, the frequency at which the optical depth is unity is expected to be 46 MHz. Packolczyk gives tables for calculating the frequency v_{max} at which the spectrum will peak. After applying factors for redshift which correct for both the intrinsic surface brightness of the source and the effective frequency of the observations, the predicted frequency $v_{max} = 73$ MHz. This is very close to the actual peak in the spectrum in Fig 2. However the agreement can be taken as no more than an indication that the

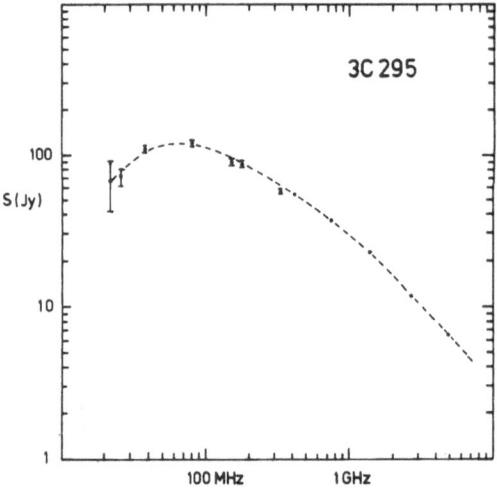

Fig 2   Radio spectrum of the total flux of 3C295.

filling factor is close to unity, since there is a wide range of surface brightness over the source and a corresponding range of v_{max} from about 30 MHz to over 100 MHz is different parts. Even this crude analysis excludes the possibility of extreme filling factors as low as one per cent.

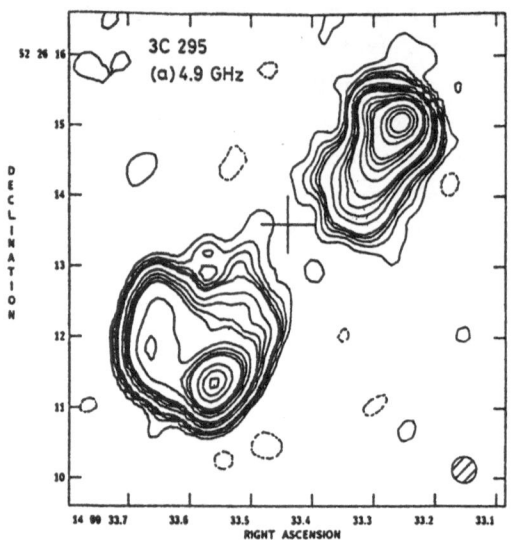

Fig 3   4.9 GHz map of 3C295 with 0.3 arcsec resolution from VLA observations. Contours are at 2, 4, 6, 8, 10, 20, 30, 40, 50, 100, 150, 200, 250, 500, 750, 1000 mJy/beam.

The application of the same method to sources with much lower surface brightness at low frequencies is straighforward. Objects with surface brightness about 10^5 times fainter ($\sim$ 1mJy in a 3 arcsec beam) will have values of v_{max} about 100 times lower, i.e. about 1 MHz. These brightness are typical of the bridges of double radio source. Fig 4 shows the intrinsic radio luminosities and overall physical sizes of the sources in the Laing, Riley and Longair 3C sample. On the crude assumption that most of the flux at low frequencies arises from a uniform cylindrical bridge with an axial ratio of 5, the expected values of v_{max} for a filling factor of unity are plotted on the figure. All except the very largest and smallest sources have values lying in the range 1-30 MHz which are precisely those frequencies that concern us in this meeting.

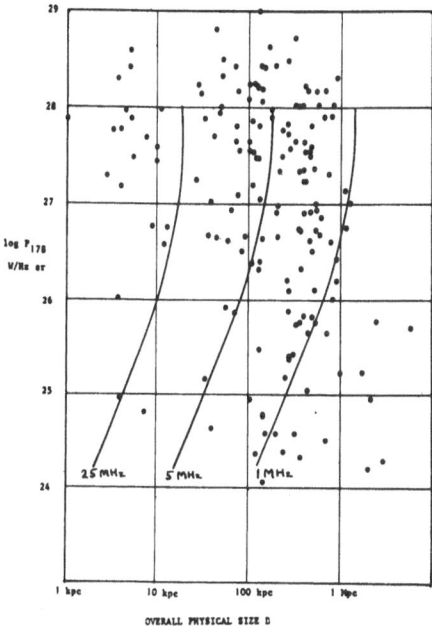

Fig 4   Radio luminosity and overall physical size for the LRL sample, showing the frequencies at which the bridges and extended lobes of the sources are expected to show synchrotron self-absorption for a synchrotron emission filling factor of unity.

I think it would be of great interest to try to determine the filling factors for synchrotron radiation in sources but I do not pretend that it is an easy programme. It would naturally require mapping the sources at frequencies of 20 - 30 MHz with arcsecond resolution. But this is what we would like to do in any case for studies of the spectral variations across sources in connection with their dynamics and life histories. The next critical question to address is whether this is feasible or not.

5.   The design of high resolution ground-based array at low frequencies

In section 2 we saw that it was unlikely that observations above 10 MHz could be carried out from space, except perhaps from the far side of the Moon. Here we look at what can be done from the ground.

The main limitation is the ionosphere. At the lower frequencies observation at night will be essential; only at night is the ionospheric critical frequency sufficently low that interference from long-distance propagation of man made signals is absent on good sites. The effects of the ionosphere on low frequency observations are then dominated by travelling waves with wavelengths of typically 200 km, speeds of ~ 150 m/s, and corresponding periods of about 20 mins. For radio telescopes up to a few km in size they behave as moving prisms whose only effect is to deflect the line of sight to a source through small angles. For an interferometer this is usually expressed as a small difference in phase path to the two interferometer elements which is just proportional to the separation of the elements. Fig 5 gives statistics of these phase fluctuations at Cambridge, which may be typical of

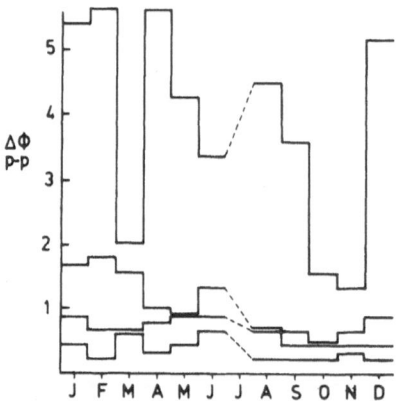

Fig 5   Monthly statistics of ionospheric phase fluctuations in radians at 151 MHz on a 4 km baseline at Cambridge. The four histograms are the maximum phase deviations during an observation of 4-12^h duration for the worst day, the median day, the median night and the best night in each month in 1984.

mid-latitude sites, in 1984. Experience suggests that the best conditions occur at night-time, in winter, at times close to sunspot cycle minimum. The phase fluctuations are then typically 2 deg of phase/km at 151 MHz and 10 deg of phase/km at 38 MHz. In mid-cycle the conditions are maybe a factor of two worse. These values correspond to a total phase path variation across the 200 km wavelength, or during the 20 min period, of ±400° at 30 MHz or ±1200° at 10 MHz. These obviously require accurate correction in any large proposed array.

The method of correction is straightforward in principle. The sky is well mapped with resolutions of one to a few arcmin at, for instance, 151 MHz with the 6C and 7C surveys. The positions of sources of ~ 1 Jy are known at this frequency to an accuracy of ~ 1 arcsec and have a surface density on the sky of ~ 1 deg^{-2}. These provide a grid of reference sources which can be used to calibrate each part of a large array. For instance, a 10 km section of a 10 MHz array has a resolution of 12 arcmin, sufficient to distinguish such a source in a particular direction and use it to determine the instantaneous phase slope across the 10 km section. Curvature of the wavefront due to the ionosphere across this section is only ~ 15° so that there is no serious distortion of the beam. In this way a very large array can be calibrated piecemeal so that high resolution imaging can be achieved within an area for which the corrections are valid, i.e. within the isoplanatic area. The radius of the isoplanatic area is > 1° for the typical conditions described here even at 10 MHz. So it seems to me that there is no doubt that the limiting useful resolutions of 1 arcsec at 30 MHz and

10 arcsec at 10 MHz set by interstellar scintillation can be achieved with certainty with ground-based arrays.

The choice of the site will be crucial. At present there is scarcely any published data which enable an informed choice to be made. Such investigations are not expensive and should be carried out now.

References

Baldwin, J.E., Boysen, R.C., Hales, S.E.G., Jennings, J.E., Waggett, P.C., Warner, P.J. and Wilson, D.M.A., 1985. Mon. Not. R. astr. Soc., 217, 717.
Bridle, A.H. and Purton, C.R., 1968. Astron. J. 73, 717.
Kellermann, K.I., 1964. Astrophys. J. 140, 969.
Long, R.J., 1963. Ph.D. thesis. University of Cambridge.
Packolczyk, 1970. Radio Astrophysics, Freeman, San Francisco.
Rees, N., 1990. Mon. Not. R. astr. Soc., in press.
Roger, R.S., Costain, C.H. and Lacey, J.D., 1969. Astron. J. 74, 366.
Simon, A.J.B., 1977. Mon. Not. R. astr. Soc., 180, 429.
Slee, O.B. and Higgins, C.S., 1973. Aust. J. Phys. Astrophys. Suppl. No. 27.
Slee, O.B., 1977. Aust. J. Phys. Astrophys. Suppl. No. 43.
Viner, M.R. and Erickson, W.C., 1975. Astron. J. 80. 931.
Williams, P.J.S. and Bridle, A.H., 1967. Observatory, 87, 280.
Williams, P.J.S., Kenderdine, S. and Baldwin, J.E., 1966. Mem. R. astr. Soc., 70, 53.

The new Cambridge 38 MHz radio survey – 1 steradian at 4 arcmin resolution and sub-Jansky sensitivity

Nick Rees

Mullard Radio Astronomy Observatory, University of Cambridge,
Cavendish Laboratory, Madingley Road, Cambridge, CB3 0HE, England.

ABSTRACT

A high resolution 38 MHz radio survey has been performed at MRAO in Cambridge and the results are currently in press. The survey was extremely successful despite being performed on a low budget and on a restrictive site. The design and results of this telescope must be considered when making comparisons between proposed ground-based and space-based high performance low frequency telescopes.

1 INTRODUCTION

This talk is to be totally ground-based, both from the point of view of my feet, and of the telescope I will be describing. Why, you may ask, is it being presented at a workshop entitled "Low Frequency Astrophysics from Space"? – something which definitely implies the latter and which, cynics may argue, often implies the former. The answer is that a step into space must be made only after consideration of what has been done, and what can be done, by telescopes on the ground. The relative scientific merit of the ground based option should be considered, along with its funding potential and other pertinent questions, before a final decision is made.

Where do I fit in in all this? For the past five years I have been responsible for building, operating and reducing the data from the new 38 MHz radio telescope at MRAO in Cambridge. It has performed what is arguably the best low frequency survey to date – about an order of magnitude better in resolution and sensitivity than the best previously published 38 MHz survey (see [5]) – and as such it is an instrument that should be considered in detail when making realistic judgements of the relative merits of space-based versus ground-based astronomy.

I readily accept that these claims have some drawbacks and counter arguments; for example the survey only covers a relatively restricted area ($\delta \gtrsim 60°$) and at a wavelength of 8 m it is not even truly decametric – maybe it isn't even low frequency. However this is not really the point, none of these arguments detract from the importance of considering the Cambridge telescope as a successful example of a modern, low-frequency, ground-based radio telescope.

The aim of the survey can be summarised as:

"To provide a cheap, quick, low risk, continuum survey of the region around the North Celestial Pole comparable in resolution and sensitivity to the 151 MHz 6C survey, but at a factor of four lower in frequency (i.e. 38 MHz)."

Within these words is concealed the fact that both the resolution and sensitivity were to be an order of magnitude better than the last, big, 38 MHz survey and it was to be the first survey below 100 MHz which was not going to be totally confusion limited. Both of these factors highlight the science that we hoped to obtain and in this paper I hope to be able to give some brief idea on how we achieved it.

2 DESIGN

I am not going to describe the instrument in great detail because this is done elsewhere ([3] & [4]). Instead, I will concentrate on the principles behind the design in the hope that this may provide a greater insight, and be ultimately of greater use when other people come to designing future instruments.

When designing the telescope we had two main constraints, both of which would be relaxed for a future telescope.

The site. The telescope had to fit on the Cambridge 5 km baseline. This meant that the baseline could not be due East-West and that there was a limited choice of aerial positions both because of topographical constraints, and the need not to interfere with the two other telescopes already on the baseline. The small average width of the baseline also restricted the number of pointing directions for which the aerials could be effectively guyed.

The budget. The limited budget of £50,000 (1985) meant that maximum use had to be made of existing materials and, in particular, the electronics had to interface with the existing 151 MHz Cambridge Low-Frequency Synthesis Telescope (CLFST) electronics at the IF stage. This restricted the choice of aerial positions and further limited us to a maximum of 60 aerials, and 784 of the 1770 total possible complex correlations. The limited budget also meant that the aerials could not track, and so the survey had to be synthesised out of a series of discrete pointings in a similar way to the 6C survey (see [1] and [2]).

Additionally, the telescope was designed purely as a single-frequency, short-term survey instrument and I feel that this was the major philosophical difference between it and the Clarke Lake TPT. This simplicity of purpose saved time and money and in I feel that it contributed to the telescope's ultimate success.

The design is illustrated in fig. 1 and it represents a departure from the traditional decametric (and near-decametric) design of a T-shaped array with a large number of uniformly-spaced aerial elements. The new design, which is necessarily similar to that of the CLFST, has a lot more in common with high frequency telescopes such as Westerbork and the Ryle (ex Cambridge 5 km) than with the traditional low-frequency design. It is a linear, East-West array with the individual aerial elements being positioned in a relatively sparse arrangement such that the correlated aerial pairs form a uniform radial pattern in the aperture plane. The aerials themselves are 10 element Yagis some 18 m

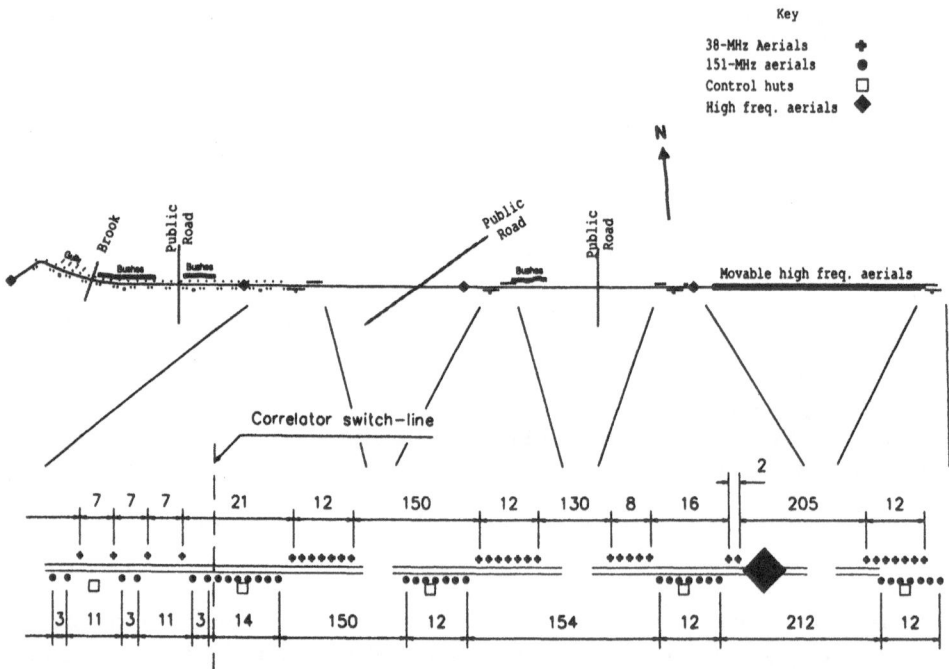

Figure 1: The layout of the Cambridge 5-km baseline. The upper part of the diagram is approximately to scale and shows the positions of all the aerials on the baseline. The baseline is so narrow that very little overlap can be tolerated. The lower half of the diagram is an expanded view of the regions of the baseline occupied by low-frequency aerials and it shows the distances between the low frequency aerials in 'spacing units' (approximately 0.75λ at 38 MHz). 26 aerials of each frequency are omitted to the West (left) of the picture, but these follow the same regular (7,7,7,7) or (3,11,3,11) pattern that is shown. Aerials to the East of the correlator switch-line are only correlated against those to the West of the line. The result is a very uniform aperture plane coverage with the correlated baselines at multiples of the spacing unit.

long, and so they have a higher gain than most other aerials that have been used at these frequencies.

The spacing arrangement is also best understood by comparing it with the high frequency telescopes. These generally have a number of aerials on a railway track which can used to synthesize a continuous array of aerials, correlated against a set of uniformly spaced outlying aerials. Our arrangement is basically the same except that because aerials are so cheap we do not have to synthesize the continuous section, we have got just a continuous line of aerials, and the outlying aerials have been replaced by groups of aerials to increase sensitivity and to increase the visibility density in the aperture plane. This results in $\gg 1$ visibility (aperture gridpoint)$^{-1}$, a very important feature since it, combined with the East-West baseline, gives a very 'clean' synthesised beamshape which means that the maps do not have to be deconvolved.

The exact configuration of a telescope of this sort is largely determined by two factors.

1. The maximum baseline. Clearly, this determines the maximum resolution of the telescope but it also governs the size of the isoplanatic patch and so, indirectly, it determines a minimum collecting area for the telescope. This collecting area must be enough to be able to calibrate the ionospheric effects and so the telescope's sensitivity must be sufficient to detect, with reasonable signal to noise, at least one source within the isoplanatic patch within an integration time in which the ionosphere is relatively stable.

2. The size of the primary beam of the individual antenna elements. Since the total collecting area is already determined, this then determines the number of aerials needed for the required sensitivity. However, it also determines the maximum unit spacing of the pattern of correlated aerial pairs, because this must be small enough that sources within the primary beam have grating rings that lie outside the primary beam.

For our telescope the first of these was fixed at 4.6 km and the aerials we used had a FWHM of about 35°. This enabled us to calibrate on an $\approx$ 20 Jy source at the centre of the primary beam, of which which there are about 50 in the survey area. In practice the number of useable sources was less than this because, of course, the sources were not at the centre of the primary beam and any case they drifted through (and out) of the beam as the aerials did not track.

3   DATA REDUCTION – PROBLEMS AND SOLUTIONS

The survey was produced from the thirty nine best 24 h observations made between June 1986 and June 1987. The processing can be divided into three major stages – visibility data processing, map synthesis and map plane processing.

3.1   Visibility data processing

The major problems to be overcome during the visibility data processing were the removal of the phase distortions introduced by the ionosphere, and the removal of the effects of Cas A and Cyg A. These two sources dominated the visibility data for much of each 24-h observation (see fig. 2) and their side-lobes are detectable over most of the survey region. The methods we used to cope with these problems were fairly simple. Calibrating out the ionosphere was done by observing the apparent position shift of a bright source in the vicinity of the field and applying the calculated phase gradient to the visibilities. The flux removal was essentially as simple – subtracting a coherent time average of the observed visibility data, except things were slightly complicated by the systematic amplitude variations caused by the source transitting the primary beam. Both these procedures are explained in greater detail elsewhere, but it is enough to say that, although simple, the procedures performed surprisingly well (see fig. 3).

What about interference, the other scourge of low frequency astronomy? This, in fact, proved no problem at all and this was as much a surprise to me as anyone. Whether this was due to it being the solar minimum, transmitters being moved up frequency for larger bandwidth, 38 MHz being a frequency high enough that long distance propagation

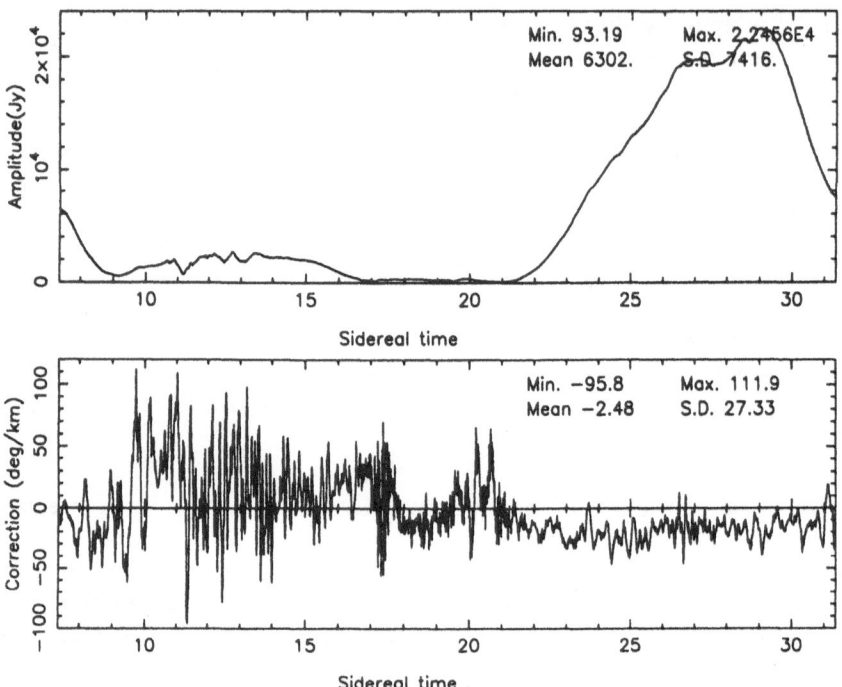

Figure 2: The effect of Cas A on a typical observation. The top picture shows the variation in the apparent flux of Cas A seen by the telescope. The modulation is caused by the primary beam, and to a lesser extent, by the changing sky temperature affecting the AGC of the front-ends. The total flux of Cas A is 32,000 Jy, which should be compared with the total flux in the primary beam from all other sources (≈ 500 Jy), and the noise on the final survey maps (≈ 250 mJy). The bottom picture is from the same observation and shows the changing phase slope across the baseline caused by the ionosphere in the direction of Cas A. The slope was usually linear over the 5 km baseline of the telescope.

is largely eliminated or any other reason, the only major interference problem we had was self generated and easily solved. For this, I must say, I am grateful.

Nevertheless, I must admit that all did not go quite according to plan and we could have done with slightly more bright ionospheric calibration sources in the sky. This was really a problem of trying to get as much resolution as possible for the money – inevitably we did not have quite as much collecting area as we would have wished and so the calibration was not as good as we would have wished. In spite of this, I don't think the accuracy of the source flux densities was compromised much, so from the point of view of the survey the problems were not important.

3.2 Map synthesis

Map synthesis was fairly straightforward with only a few minor twists. Optimal gridding and grading (taper) functions were used to reduce aliases and sidelobes respectively. Additionally, data samples which had a high variance were gridded with relatively less

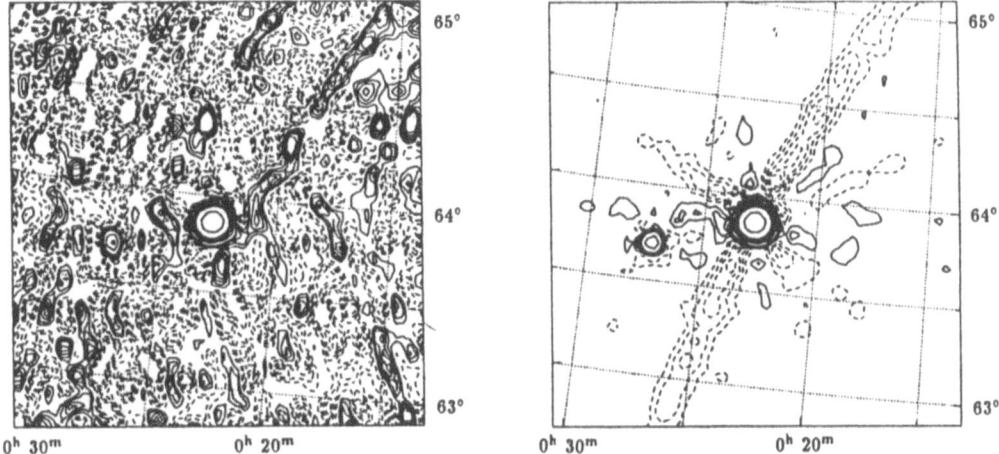

Figure 3: The effect of the source removals on 3C10. The left hand picture is a map made without Cas A or Cyg A removed, and the right hand one is one has both sources removed. The increase in the map dynamic range is obvious and the measured flux of the source on the two maps differs by less than 2%. Contour levels are nominally at:
-10, -8, -6, -4, -2, 2, 4, 6, 8, 10, 20, 30, 40, 50, 100 Jy/beam.

weight since this improved the final map quality without the side effects that occur if the data is rejected completely.

3.3 Map plane processing

Calibrating the survey was the biggest problem that had to be overcome in the map plane. The first step was to ensure that the flux scale was consistent over the whole survey area – no easy task when the beam-shape of your aerials varies greatly as a function of pointing and you have no way of measuring this variation. One obvious way of dealing with this was to compare the fluxes with other surveys. I tried this and concluded that there were no other surveys that were good enough for the job – at least this gave me the reassurance that what I was doing was was worthwhile. Eventually, I decided that the only way to determine the flux scale variations was through internal consistency checks – transit scans, comparisons of source counts in different survey regions and comparisons of the apparent flux of the sources which appeared on more than one map.

The results of this exercise seem to be quite encouraging. At the time I did it I felt that the accuracy of the procedure was probably about 10 per cent and I recommended that this figure be adopted, but recent spectral index comparisons with new data from the 6C survey indicate that the errors could be as low as 5 per cent – almost too good to hope for.

The second stage of the calibration was to tie the fluxes onto an absolute flux scale and, after a considerable amount of thought, I decided to adopt the scale of Roger, Bridle and Costain since the Baars scale is confused by the flux variations of Cas A (see [4]).

After the calibration, the source fitting routines were fairly simple. One minor problem was that the height and shape of the true synthesised beam varied due to ionospheric effects and so 'point' source fitting routines appeared to underestimate the flux of sources by up to 25 per cent. Fortunately, the total flux of the source was unchanged and so the difficulties were resolved by analysing every source as if it were resolved.

4   WHAT CAN WE GET FROM THE SURVEY?

Work is already being done on the astronomical side by a few students and their supervisors at MRAO, and members of George Miley's group at Leiden. It is basically an investigation of all the classical talking points of low frequency astronomy, but the area of interest has now been defined because of the difficulty of obtaining more 38 MHz information in the immediate future. The first thing they are trying to do is get together as many existing radio catalogues (published and unpublished) as they can get their hands on, just to see what has already been done.

After this is completed the next step is to observe the gaps with the CLFST at 151 MHz, Westerbork at 328 and 610 MHz and the VLA at higher frequencies. They then will go through the normal procedure about trying to pick up information in other wavebands, and the aim is to concentrate on the steep spectrum objects first. These may be (because of the $P - \alpha$ correlation) high P, high z sources, or they may be near-dead radio sources in clusters of galaxies. For the latter we are fortunate the the 38 MHz survey includes the North Ecliptic Pole, the deep area of ROSAT.

The other item of interest that could be gleaned from the survey is a detailed prediction of the viability of future ground based, low frequency instruments. I must admit we followed the procedure of not investigating this closely unless it was absolutely necessary, and since it wasn't, we didn't. This reflects that we are radio astronomers, not ionospheric scientists, and so what may be needed is a collaboration between the two fields.

5   WHAT NEXT?

I suppose one of the crucial questions is whether we can build a ground based array which could match the performance of a space array above about 20 MHz. Clearly this would be a challenge, but if we could we would have to consider the economics of both proposals before committing ourselves to one or the other. One can put a great deal of hardware on the ground before you even approach the cost of putting a small amount of hardware into space.

Such a ground based instrument would necessarily comprise of a number of compact arrays on a baseline of over 1000 km. Each compact array would be the equivalent of one station in a VLBI array and would have a collecting area large enough to be able to detect a ≈ 1 Jy calibration source in an integration time of a minute or so. It would also seem wise that the dimensions of each compact array would be small enough to require only a single, linear, ionospheric phase gradient correction in the integration time.

After establishing this general picture there would be a number of practical questions

Figure 4: The 38MHz survey. The North Celestial Pole is at the centre of the picture and RA 0^{hr} is at the top. The minimum declination is about 55° and the survey is complete to declination 60°. The brightest source is Cas A, which is 32,000 Jy and dominates the top left of the picture. The faintest source is less than 1 Jy and thus the overall dynamic range is about 40,000:1.

still to answer:

- Should all elements of the telescope be correlated, or should the compact arrays just be actively phased?

- Should correlation be done during the observation or after?

- What bandwidth should be used? Falling electronics costs may mean that a wide bandwidth, multichannel, instrument may be a better way of achieving sensitivity (and maybe $u - v$ coverage?) than having more aerial elements. A large bandwidth may help calibration since the ionospheric phase delay is a fairly fixed function of frequency, but a lot of channels would be needed for a fractional bandwidth that could approach 0.5. Or do we just digitize at 80 MHz and sort it all out afterwards? Can we?

- Exactly what aerial design should be used? Should the aerials track, and if so electronically or mechanically?

- Where should it be sited? Tasmania seems to be a widely held ideal, but knowing the place as I do, obtaining sites would require a lot of planning, and it possibly isn't big enough.

- Where does the money and manpower come from?

I must admit to not knowing the answers to any of these questions but I do feel that there could be more spinoffs from such an instrument than from a space based one. This is not to say that it would be a high risk exercise. On the contrary, it could be built in an incremental way, making an instrument which was, say, ten times the resolution and sensitivity of the current 38 MHz survey for one solar minima, with the aim of extending it, if successful, for the next solar minima (provided, of course, we can cope with a 15-20 year planning cycle). On the other hand, whilst a space telescope is a relatively low technological risk, it does suffer from a risk of catastrophic failure.

What of the technological risk of a ground-based telescope? It really boils down to what computing power will we have available and what will the current bunch of calibration algorithms look like. I feel I should point out to people that at the time of the last, big, 38 MHz survey the first of the modern synthesis instruments (the Cambridge One-Mile) had not been built, we were using EDSAC to do our fourier transforms, Fortran had been released a few years ago and VLBI was thought to be impossible. However, we had already sent satellites into space and it was, as it still is, painfully expensive.

I must quickly add that I am not arguing against a space based platform. What I am saying is that our best survey to date between 20 and 50 MHz had a direct cost of $\approx$ \$80,000. I do not quite see the logic in spending about 1000 times this on space based telescope without doing a little bit more ground-work first. As in many academic pursuits, it is right to speculate at any time, but we should temper our enthusiasm when it comes to spending money on actual hardware.

6 REFERENCES

[1] Baldwin, J.E., Boysen, R.C., Hales, S.E.G., Jennings, J.E., Waggett, P.C., Warner, P.J. & Wilson, D.M.A., 1985. *Mon. Not. R. astr. Soc.,* **217**, 717.

[2] Hales, S.E.G., Baldwin, J.E. & Warner, P.J., 1988. *Mon. Not. R. astr. Soc.,* **234**, 919.

[3] Rees, Nick, 1989. *Ph.D. Thesis,* Cambridge University.

[4] Rees, Nick, 1990. *Mon. Not. R. astr. Soc.,* (in press – two papers).

[5] Williams, P.J.S., Kenderdine, S. & Baldwin, J.E., 1966. *Mem. R. astr. Soc.,* **70**, 53.

VII. COSMIC RAYS AND THE GALACTIC BACKGROUND

COSMIC RAYS IN THE GALAXY AND THEIR IMPLICATIONS FOR VLF RADIO ASTRONOMY

W.R. Webber
Astronomy Department
New Mexico State University
Las Cruces, NM 88003

INTRODUCTION

We shall begin with a few new observations that, while seemingly unrelated to the subject of VLF Radio Astronomy, are useful for placing in perspective the types of measurements that will be important to make in this field.

First, there are new measurements of the local galactic cosmic ray (GCR) energy density based on improved solar modulation estimates based on interplanetary gradient observations from the Pioneer and Voyager spacecraft that give $E_{cr}> 1.5$ eVcm^{-3}, a factor of perhaps 2 higher than the commonly adopted value. (Webber. 1987) This corresponds to $> 2.5 \times 10^{-12}$ erg cm^{-3} and if we compare this with the magnetic field energy density, $B^2/8\pi$, requires an equipartition $B>8\mu G$, considerably higher than the usually adopted values of 3-5 μG for the average interstellar field. For uniform cosmic ray diffusion in the galaxy it is usually assumed that E_{cr} has to be $<E_B$, so one is immediately faced with the question, does the field really control GCR or visa-versa? Indeed perhaps a high GCR energy density in the disk of the galaxy is responsible for blowing out this field to form an extended disk halo along with a convective wind, such as may be observed in radio maps of other nearby spiral galaxies. Alternatively, perhaps the GCR energy density is unusually high at the sun. Other evidence on this and on fluctuations in the B field strength that may help understand these inconsistencies, and the way in which VLF radio observations can be used to shed light on this situation will be discussed shortly.

The second new observation relates to the distribution of GCR nuclei in the galaxy as deduced from a study of -rays of E> 100MeV, from COS-B (Bloeman, 1989). These γ-rays arise mainly from π^0 decay, with the π^0 being produced by nuclear interactions of GCR with the ambient interstellar H, both HII and HI. The observed γ-ray distribution in the galactic disk is thus $\sim \int n_{cr} \cdot n_H \cdot dl$ along a line of sight. Greatly improved information on n_H, both HII and HI has allowed Bloeman to interpret the γ-ray data in terms of the radial distribution of cosmic ray nuclei, n_{cr}, in the galaxy. This distribution is shown as a series of solid points in Figure 1. The weak radial dependence of this distribution derived from >100 MeV rays is evident.

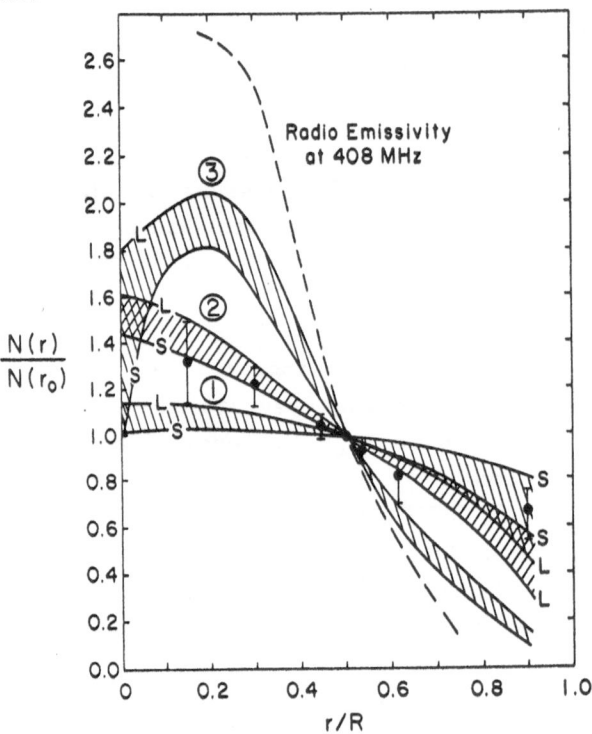

Figure 1. Caculated and deduced radial profiles of GCR in the galaxy (Bloeman, 1989). Shaded areas include radial profiles for halo thickness between 1Kpc(S) and 10 Kpc.(L) Source radial distributions considered are (1) flat, (2) exponential with r_0=20Kpc, (3) supernova distribution.

Recently we have attempted to model the expected radial distribution of cosmic rays for various possible source distributions for realistic propagation models including a halo + thin matter disk (Gupta, Lee, and Webber. 1989). The results of this modelling are also shown in Figure 1. It is observed that to obtain the relatively flat distribution of n_{cr} that Bloeman deduces requires a relatively flat source distribution. Radial distributions calculated for a thin halo

(S) of 1 Kpc thickness and a thick halo (L) of 10 Kpc thickness are shown for:

1) A flat source distribution,
2) One that has an exponential decrease with $r_o=20Kpc$
3) One that follows the SNR radial distribution in the galaxy.

An important new aspect of this study was to use the observed Be/C ratio in cosmic rays and the measured fraction (~0.25) of the radioactive isotope ^{10}Be remaining in cosmic ray to determine that the thickness of the cosmic ray halo must be <~ 2-3 Kpc. Thus thick halos cannot be invoked to "flatten" the radial distribution of n_{cr} expected from SNR. It is clear, therefore, that the distribution of n_{cr} obtained from γ-rays requires a relatively flat source distribution which is inconsistent with the observed distribution of SNR.

At this point we should note that γ-rays < 70 MeV originate primarily from cosmic ray electron bremsstrahling with the interstellar gas, in contrast to the higher energies which arise primarily from π^o decay. These cosmic ray electrons, which have energies < 200 MeV, exhibit a very similar radial profile to the higher energy nuclei, as deduced from the < 70 MeV rays by Bloeman.

In general it is believed that the ratio E_{el}/E_{cr} is ~ a few percent for galactic cosmic rays, however VLF radio astronomy in conjunction with measurements of the lower energy γ-rays can help answer several important questions regarding this ratio and its variability.

1) Does it vary substantially in our own galaxy? This might be indicative of different sources for electrons and nuclei.
2) Is this ratio different in different galaxies where both radio and γ-ray observations can be made, eg LMC, M-31, various AGN?
3) Are the features of the electron distributions in galaxies as deduced from the measurements of radio halos consistent with what is expected for the propagation of cosmic ray nuclei?
4) How can features of the IS medium and the distribution of cosmic ray electrons as deduced from VLF radio measurements be used to understand the acceleration and propagation of cosmic rays as a whole in our galaxy.

VLF Radio Astronomy and Cosmic Ray Electrons

To begin this discussion we show a 30MHZ radio map of our galaxy as produced by Cane, 1979.

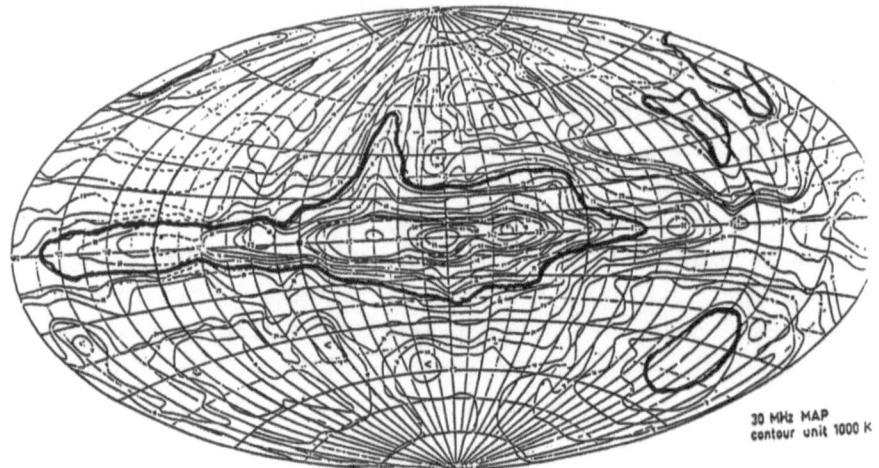

30 MHz MAP
contour unit 1000 K

Figure 2. Radio map of galaxy at 30MHZ from Cane, 1979

This map has an angular resolution ~7°. The heavy line represents a contour of constant brightness temperature. This is a map of moderate angular resolution at the upper frequency end of what we consider to be the VLF band. The following comments can be made in reference to this map.

1) This map looks very much like maps of other nearby spiral galaxies that have been made at much higher frequencies - specifically the feature of a thick radio disk.

2) There is very little indication of interstellar absorption at this frequency and for this resolution. One needs higher resolution <1° and lower frequency to see local absorption (or emission) features such as HII regions or SNR.

3) No external galaxies are evident, for example LMC, M-31 or AGN. Again one needs resolutions < 1° and then it will be possible to compare the radio emmission directly with both low and high energy γ-ray measurements of the same galaxies.

4) There are large regions of minimum sky brightness. What is the low frequency cut-off in these regions and are these windows out of the galaxy? What part of this emission is truly EG in nature? Relevant to this question is the escape of cosmic ray electrons from galaxies and the strength of IG magnetic fields.

To discuss in more detail VLF radio astronomy and its interpretation in terms of cosmic ray electrons we need to recall a few important relations. For a distribution of electrons with spectral index q=-2.5 the peak radio emission will occur at a frequency

$$\gamma \; (\text{MHZ}) = (5\text{--}10)\, B^2\,(\mu G) \cdot E^2\,(\text{GeV})$$

So for γ = 30 MHZ and B = 6μG, E~0.7 GeV thus VLF radio astronomy in our galaxy is, in general, a study of low energy electrons. Also:

emissivity = $\mathbf{\mathcal{E}}$ = $\mathbf{n_{el} \cdot B}$ $\mathbf{(q+1)/2}$

spectral index = q = 2$\boldsymbol\alpha$ +1

lifetime = $\mathbf{t_{el}}$ = $\mathbf{B^{-1.5} \cdot E^{-1}}$

The spectral shape of the "local" galactic electron spectrum is best derived from shape of the polar radio spectrum (Figure 3) where the radio index α=~ 0.62 implying q = -2.24.

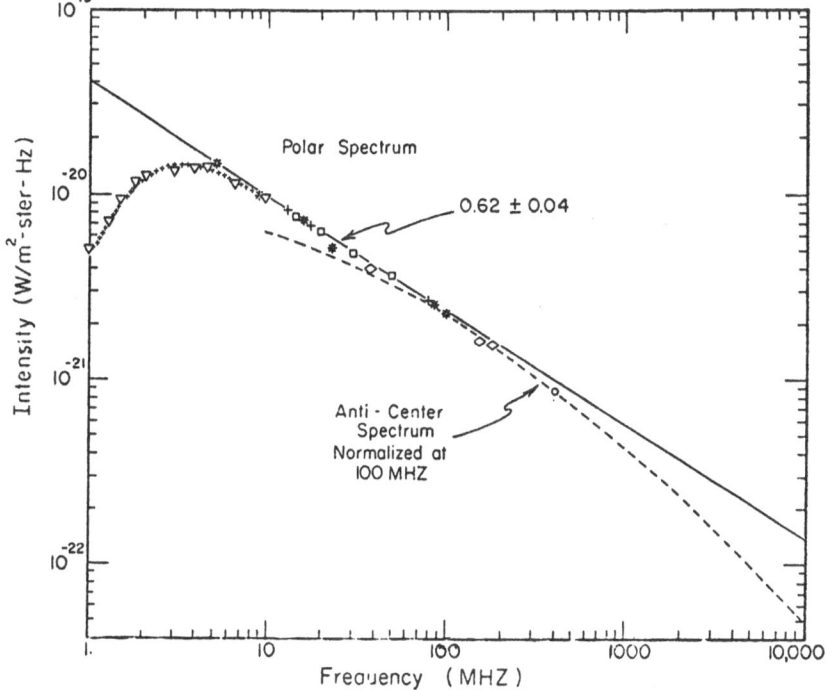

Figure 3. Spectrum of radio emission in the polar direction

This is basically an electron "source" spectrum and it is consistent with the deduced source spectrum for cosmic ray nuclei. This equality is an important fact that relates to the acceleration of these two types of particles. If one assumes that this same electron spectrum extends to lower energies as predicted by propagation calculations, then the low frequency turnover in the polar radio spectrum can be used to estimate the amount of free - free absorption and from the detailed shape of this turnover, to estimate the relative amount of emission coming from the disk where there is mixed emission and absorption and that coming from an extended halo or from outside the galaxy which is absorbed exponentially. These types of estimates lead to:

1)   The absorption depth Υ ~1 at 2 MHZ. This, taken with an effective disk half thickness of 1 Kpc, leads to an average electron density n_e~0.025cm^{-3}

2) The total amount of emission from the disk + halo is ~ 88% and the EG emission ~12% of the total polar emission.

Typical estimates place the emissivity of the halo at ~0.04 of that for the disk at 10 MHZ, for a halo of half thickness~6Kpc. The corresponding emissivity of the disk is thus 4 x 10^{-40}Wm^{-3}HZ^{-1}sr^{-1}. These values, particularly for the halo and EG components, have considerable uncertainty due in large part to the very crude angular resolution inherent in the radio measurements. Improved angular resolution measurements, of say a few degrees at ~1-2MHZ, could begin to examine detailed structure in both emission and absorption and greatly improve our understanding of the points discussed above.

Using this polar radio spectrum and the (local) emissivity as a function of frequency derived from it is an important first step in a comparison with the measured electron spectrum at earth. A difficulty in this comparison is solar modulation, which is still of uncertain overall magnitude, but probably decreases the electron intensity at earth below ~1 GeV by an energy dependent factor of at least 2 below that observed in nearby IS space. One can attempt to use the polar radio spectrum as a template, normalizing the implied electron spectrum from this radio spectrum and the observed electron spectrum at ~10 GeV where the solar modulation is small. If this is done the free variable in this normalization, the interstellar magnetic field, is required to be at least 6-8 µG! This type of normalization is illustrated in Figure 4.

The largest uncertainty in this comparison is the exact shape of the radio spectrum above ~1GHZ since 10GeV electrons correspond to radio emission at several GHZ. In this picture the large difference between the observed electron spectrum at low energies and that inferred from the radio spectrum (a factor or nearly 100 at 0.1 GeV) must be due to solar modulation. This method of normalization

requires a large interstellar B field as is also implied by the high
local IS total cosmic ray energy density.

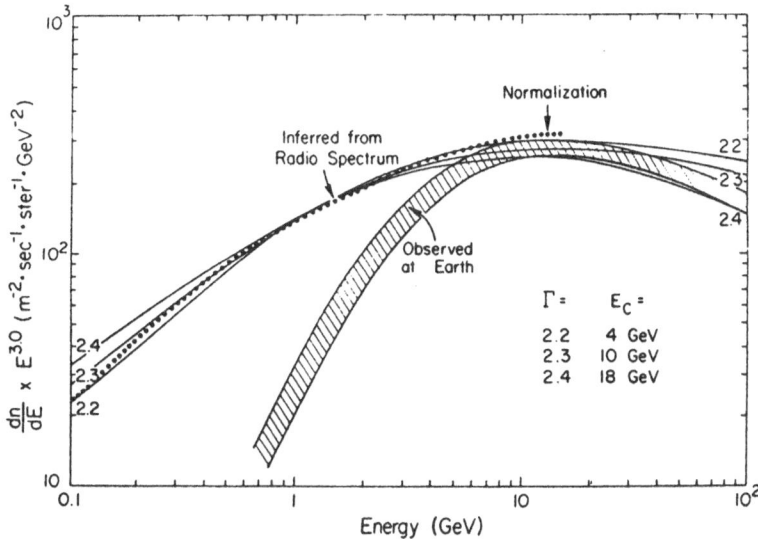

Figure 4. Comparison of measured electron spectrum at earth and
that inferred from the polar radio spectrum. Absolute normalization at
10GeV (~4-8GHZ) requires an average B~8μG.

Another way of looking at this problem is in terms of the
comparitive local emissivities as deduced from the radio data and also
from the observed local electron spectrum at earth. This data is shown
in Figure 5 where again it is seen that emissivities deduced from the
observed electron spectrum at earth are inadequate to explain those
deduced from the radio data even allowing for solar modulation unless
the interstellar field is made quite large.

The radio emissivities are derived from several different
approaches which lead to roughly the same values within a factor ~2.

One particularly useful method of determining this emissivity is
to measure the total radio emission in front of HII regions. These
regions are opaque at frequencies < 30 MHZ. To utilize this approach
requires an angular resolution < 1-2° to effectively resolve these
regions. Several workers have utilized this approach. Here we note the
work of Rockstroh and Webber, 1977, who examined ~ 15 such HII regions
in our local arm and nearby local arms using data from the highest
resolution maps available. This data is illustrated in Figure 6.

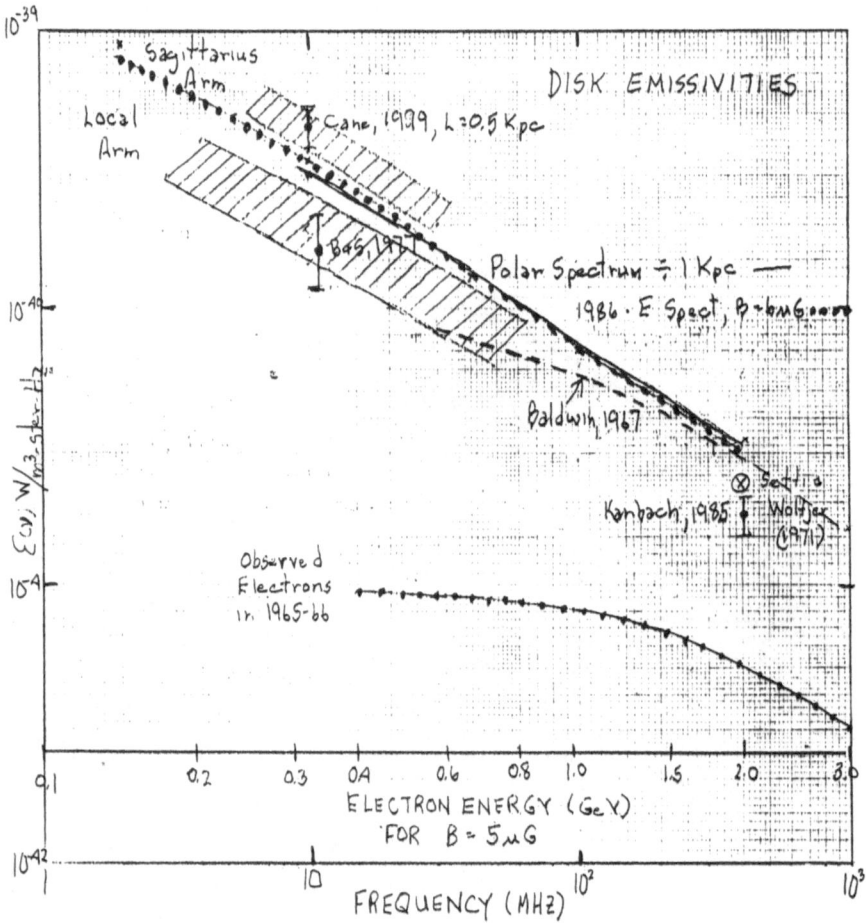

Figure 5. Comparison of local disk radio emissivities derived from radio data and from electrons observed at earth.

The average value of local emissivity from these studies agrees well with that derived from the integrated polar emission, for example, and in addition, comparing emission from different spiral arms gives evidence on the distribution of this emissivity in the galaxy. This data provides evidence for a radial fall off in emissivity with an $r_0 \sim 4$ Kpc. This is the same value of r_0 derived by Beuermann et.al., 1985, from large scale studies of radio emission at higher frequencies. Note also that this emissivity is the product of $n_{el} \cdot B^{1.7}$, so to obtain information on the radial dependence of n_{el} to compare with corresponding γ-ray data for cosmic ray nuclei one must make assumptions regarding B. Overall the problem still persists, however, that either the B field is much larger than usually assumed, or that the cosmic ray electron intensity locally is much less than the average throughout the galaxy. This latter possibility is unlikely since the local cosmic ray nuclei intensity is so large as to also require a large magnetic field. One possibility to provide for increased emissivity from a given distribution of electrons is to assume a broad distribution of B about some average value - with a tail extending to high values of B, as suggested by several workers. In this way a field with B~5µG can provide the emissivity corresponding to a much larger equivalent B, due to the $B^{1.7}$ dependence for the emissivity.

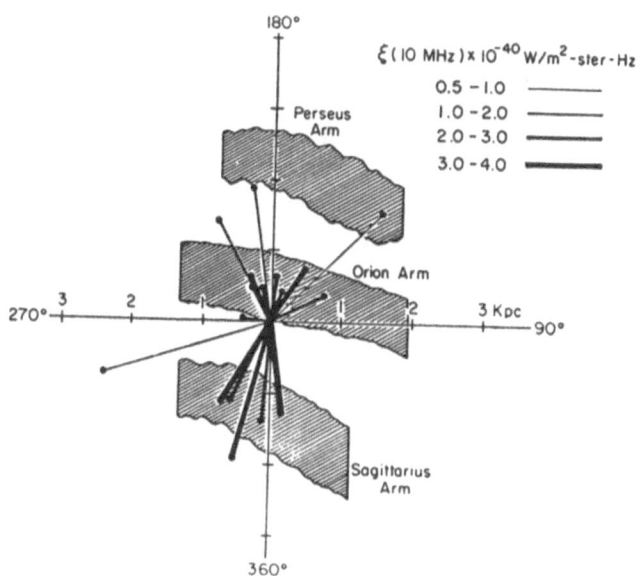

Figure 6. Line of sight radio emissivities at 10 MHZ derived from studies of HII regions (Rockstroch and Webber, 1977)

SUMMARY

In the previous sections we have summarized some of the key aspects of one important population of energetic cosmic rays with some specific details relating to these cosmic ray electrons and their observability using VLF radio astronomy. This discussion has indicated the importance of: (1) Higher angular resolution studies in the frequency range 1-30 MHZ with 1-2° resolution at 1MHZ being a useful goal. Some examples of improvements in our understanding of basic physical processes that could be made with such resolution include, (A) Further studies of HII regions leading to improvements in both estimates of radio emissivity from galactic cosmic ray electrons, and free-free absorption including n_e and T. The detailed structure and distribution of these quantities in our region of the galaxy could be studied. (B) The spectral dependence of absorption and emissivity related to SNR and giant molecular clouds. (C) Emission and absorption related to nearby galaxies and perhaps some AGN as well providing the same type of information on energetic particle populations in other galaxies. (D) The detailed fine structure of IS absorption including possible "windows" outside our immediate galaxy, leading to better estimates of IG emissivity - related to escape of electrons from galaxies and the IG magnetic field.

The second general area of importance is: (2) <u>Use of VLF radio astronomy and -ray astronomy comparisons for all types of energetic particle populations in different galaxies.</u> We cannot stress strongly enough that the electrons observed by radio-synchrotron emission $(\sim n_{el} \cdot B^{1.7})$ from 1-30 MHZ are the same electrons observed by nuclear bremsstrahlung $(n_{el} \cdot n_H) \gamma$-rays below ~ 70 MeV. This means that for all different kinds of populations of energetic particles it is possible to separate out the electron and nuclei components, and to also uniquely separate out the quantities n_H and B for these individual sources by studying simultaneously radio and γ-ray emission and thus to obtain unique information on the acceleration and transport of these species separately in different types of environments.

REFERENCES

Beuerman, K., Kanbach, B., and Berkhujsen, E.M., Astron and Ap., 153, 17, 1985
Bloeman, J.B.G.M., Ann. Rev. Astron. Ap., 27, 1989
Cane, H.V., M.N.R.A.S., 189, 465, 1979
Gupta, M., Lee, M.A., and Webber, W.R., Proc. 21st ICRC, 3, 341, 1990
Rockstroh, J.M. and Webber, W.R., Ap.J., 224, 677, 1978
Webber, W.R., Astron. Ap., 179, 277, 1987

Cosmic Rays and the Galactic Radio Background Emission

M.S. Longair

Royal Observatory, Blackford Hill, Edinburgh EH9 3HJ

1 Introduction

For many of us, the subject of this workshop brings back memories of long ago when we, and radio astronomy, were young. For me, it is an excellent opportunity to review the problems of relating the Galactic radio emission, the spectrum of cosmic ray electrons and observations of the γ-ray emission of the Galaxy. It is an opportune moment since I am preparing a new edition of my textbook "High Energy Astrophysics" and the section on this topic and its conclusions have had to be revised completely in the light of new observations of cosmic ray electrons which have been made since I sent the first edition to press in 1980. The story is now much more interesting and it has particular relevance to the subject of this workshop. I am particularly pleased that Dr Webber has already discussed this material from the perspective of an expert in the study of cosmic ray electrons. My contribution should be considered a supplementary tutorial to his exposition.

I will consider four topics, (1) the Galactic radio spectrum, (2) observations of cosmic ray electrons, (3) low energy γ-rays and (4) the propagation, acceleration and origin of the cosmic ray electrons.

2 The Determination of the Local Radio Emissivity of the Interstellar Medium

The precise determination of the Galactic radio spectrum and the extraction of the extragalactic radio background emission were topics which were actively pursued by several generations of research students at the Mullard Radio Astronomy Observatory in Cambridge in the 1960s and 1970s (e.g. Purton 1966, Bridle 1967, Webster 1971 and references therein). The spectra were determined using geometrically scaled antennae with rather broad beam-widths so that, although the angular resolution was not high, the sidelobe patterns were identical and so, provided there are no large variations in the Galactic spectrum over the sky, a good estimate of its spectrum could be made. Figure

1 shows Webster's determination of the radio spectrum of the Galaxy in the anticentre direction at high latitudes and in what are called the "inter-arm" regions. According to Webster, the spectral index of the radiation, α, is about 0.4 over the frequency range 10 to 200 MHz, steepening to about 0.9 at frequencies greater than 400 MHz. (I define the radio spectral index α by $I_\nu \propto \nu^{-\alpha}$). There is about a factor of 2 difference in intensity between the interarm and anticentre directions.

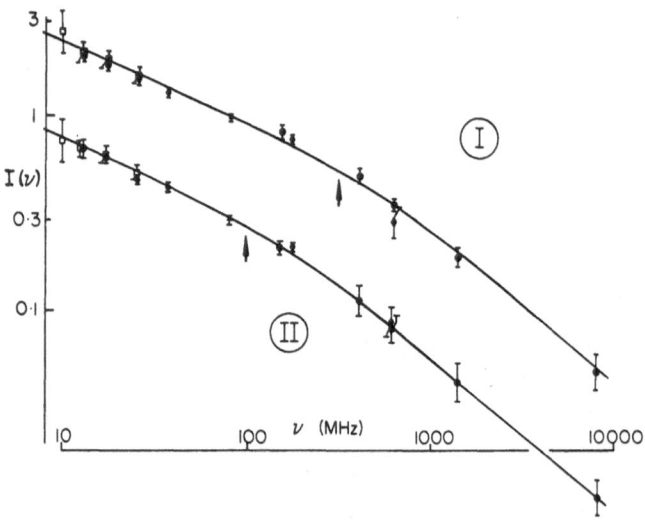

Fig. 1. The spectrum of the Galactic radio emission. Region I corresponds to the anticentre direction at high Galactic latitudes while region II corresponds to the interarm regions (Webster 1971).

The problem is then to convert this intensity spectrum into a local synchrotron emissivity for comparison with the cosmic ray electron spectrum measured at the top of the atmosphere. There are two possible approaches. In the first, an attempt is made to estimate the thickness of the radio disk from maps of the whole sky distribution of radio emission. Baldwin (1967) showed that towards the Galactic Centre, the radio emission can be described by a disk of radius 8.7 kpc with half-thickness 700 pc (in this estimate a Sun-Galactic Centre distance of 10 kpc was adopted). If this is the local thickness of the disk, the emissivity can be found. Webber (1983) adopted a half-thickness of 1 kpc. The second and, in my view, better procedure is to use optically thick HII regions as opaque screens. If the HII region is optically thick to thermal bremsstrahlung and the region completely fills the beam of the telescope, then the intensity in that direction is just the sum of the optically thick bremsstrahlung from the region plus the Galactic radio emission along the line of sight to the cloud. Knowing the thermal temperature of the cloud and its distance, the average Galactic synchrotron emissivity in that direction can be found. Caswell (1976) found an average Galactic emissivity of about 240 K kpc^{-1} which is in good agreement with the emissivity in the direction of the Galactic pole if the half-thickness of the emitting layer is assumed to be 1 kpc.

3 Comparison of the Radio Emissivity with the Local Electron Energy Distribution

Webber (1983) has surveyed observations of the cosmic ray electron energy distribution as measured at the top of the atmosphere. In his survey, he selected only observations made at sufficiently high altitudes for the effects of secondary electron production in the atmosphere to be negligible. The results of his analysis are shown in Figure 3 but, for the moment, it is only important to note that the electron energy distribution is only free from the effects of solar modulation at energies greater than 10 GeV. Below this energy, the energy spectrum is strongly attenuated and this can be entirely attributed to the fact that the electrons have to propagate to the Earth from interstellar space through the outflowing Solar Wind. Above 10 GeV, all the data can be well approximated by a spectrum of the form

$$N(E)dE = 700E^{-3.3}dE \quad \text{particles } m^{-2}s^{-1}sr^{-1}.$$

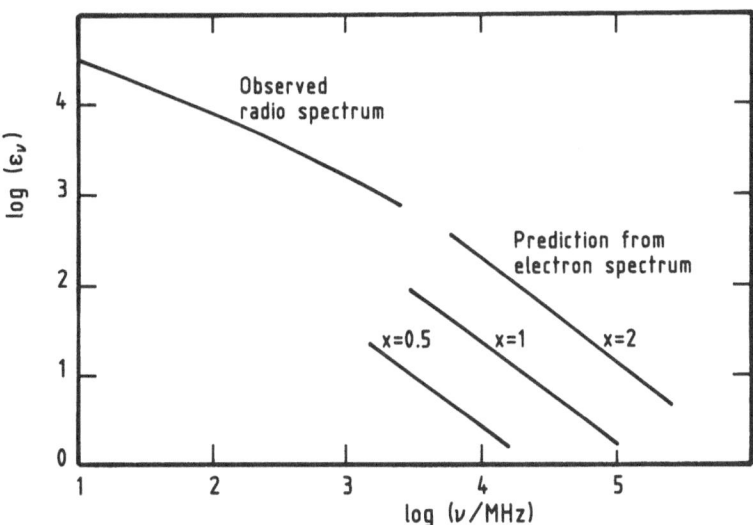

Fig. 2. Comparison of the observed radio emissivity of the interstellar medium with that expected from the local electron energy distribution for different values of the magnetic field strength. The limits of radio emissivity are relative units but the adopted emissivity at 10 MHz is $3 \times 10^{-39} \text{Wm}^{-3} \text{Hz}^{-1}$.

Assuming this flux represents the local interstellar spectrum of relativistic electrons, we can compute the local emissivity of the Galaxy assuming the magnetic field strength to be $B = 3 \times 10^{-10}x$ Tesla ($B = 3 \times 10^{-6}x$ gauss).* Figure 2 shows the result of this calculation, the low frequency end of each prediction corresponding roughly to an electron energy of 10 GeV. I have used the exact synchrotron radiation formulae for an

---

* Nowadays I use SI units as far as possible because it has become standard practice in the UK.

electron energy spectral index $a = 3.3$. Figure 2 illustrates what is an old problem. The predicted radio spectrum joins smoothly onto the observed radio spectrum, provided a high value of the interstellar magnetic field strength is adopted $B \approx 6 \times 10^{-10}$ T. It can be seen that the resulting spectrum steepens continuously from a radio spectral index $\alpha \approx 0.4$ below 200 MHz to $\alpha = (a-1)/2 = 1.15$ at frequencies greater than about 3 GHz.

There are various possible solutions to this problem:

1. We note that the radio intensity along a given line of sight is proprotional to $\int B^{\frac{a+1}{2}} dl$ and thus, if the relativistic electron distribution is uniform, greater weight is given to regions of strong magnetic field.

2. It might be that we live in a local hole in the distribution of relativistic electrons in the Galaxy.

3. Perhaps the average magnetic field in the Galaxy really is somewhat stronger than the canonical values which typically lie in the range $(1-3) \times 10^{-10}$ T (see, e.g. Rand and Kulkarni 1989, Chi and Wolfendale 1990).

In any case, the key point is that what we need to know is the variation of radio emissivity throughout the Galaxy and fortunately this is a problem which can be uniquely addressed by very low frequency radio mapping. The procedure is the same as that used by Bridle (1968), Caswell (1976) and described at this meeting by Kassim (1990). If the Galaxy is mapped at low frequencies, $\nu \sim 10$ MHz, regions of ionised hydrogen become opaque and the integrated radio emission along the line of sight to the cloud can be evaluated. The survey should be made with high angular resolution so that many opaque clouds are discovered. The key point is to ensure that the angular size of the opaque cloud is greater than the size of the telescope beam. The problem with the estimates made by Bridle (1968) and Caswell (1976) was that the observations at low frequencies did not have sufficient angular resolution or sensitivity to enable more than a few lines of sight through the Galaxy to be used. Ideally, the surveys should be carried out at a number of low radio frequencies so that the temperatures of the opaque clouds can be measured. The result is that the synchrotron emission along many sight-lines through the Galaxy can be measured. The same procedure can be used for HII regions at high latitudes. Thus, in principle, one can determine not only the average emissivity in different directions but also the fluctuations in emissivity from place to place in the Galaxy. Note that, going to frequencies lower than 10 MHz, more diffuse regions of ionised hydrogen become opaque and the volume of the Galaxy available for study decreases. This procedure combined with the types of study described by Dr Reynolds (1990) provide unique tools for studying the distribution of ionised hydrogen and relativistic electrons in the Galaxy. In my view, this is one of the most important objectives of a very low frequency radio astronomy project.

4 The Relativistic Electron Spectrum in the Local Interstellar Medium

Let us now carry out the calculation backwards so that we determine the local interstellar electron energy spectrum from astronomical observations. This is illustrated

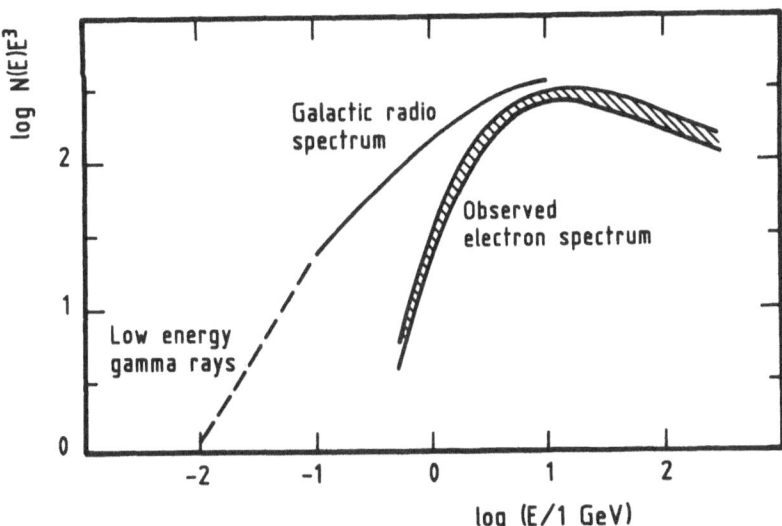

Fig. 3. The spectrum of relativistic electrons in the local interstellar medium (after Webber 1983). The observed electron energy spectrum is shown by the hatched area. The spectrum deduced from the spectrum of Galactic radio emission by a solid line. The low energy spectrum deduced from observations of low energy γ-rays is shown as a dashed line. The units of $N(E)E^3$ are particles m^{-2} s^{-1} sr^{-1} GeV^2.

in Figure 3 in which the differential electron spectrum is plotted as $E^3 N(E)$ versus E so that a spectrum $N(E) \propto E^{-3}$ would be a horizontal line (from Webber 1983). The observed spectrum of the electrons according to Webber (1983) is shown by the shaded band of observations at high energies. Joined onto that spectrum is the electron spectrum deduced from the local radio synchrotron radiation spectrum, assuming a magnetic field strength $B = 6 \times 10^{-10}$T. Note that at low radio frequencies, Webber adopted a radio spectral index $\alpha = 0.62$, which is slightly steeper than that shown in Figure 1. The large difference between the inferred electron spectrum in local interstellar space and the spectrum observed at the top of the atmosphere can be entirely attributed to the effects of solar modulation and indeed this evidence provides among the very best data on this problem.

An intriguing development has been the study of low energy γ-rays. The γ-ray spectrum at energies greater than about 35 MeV has been determined by the SAS-2 experiment and at energies greater than 75 MeV by the COS-B γ-ray satellite observatory. The spectrum in the general direction of the Galactic Centre is shown in Figure 4. The radiation consists of two components. At energies $E_\gamma \gtrsim 100$ MeV, the principal production mechanism is the decay of neutral pions π^0, which are produced in collisions between cosmic ray protons and nuclei and the interstellar gas, into two γ-rays. The spectrum of the pion decay γ-rays is well-defined and has a maximum at about 70 MeV. As seen in Figure 4, however, there is a significant flux of γ-rays with energies 40–100 MeV, in excess of that expected from pion decay. This excess can be modelled by a power-law γ-ray spectrum with photon spectral index $\alpha_\gamma = 1.6$ i.e.

$$N(E_\gamma)dE_\gamma \propto E_\gamma^{-1.6} dE_\gamma.$$

As noted by Stecker (1976) and Kniffen and Fichtel (1981), this excess can be convincingly attributed to the sum of contributions from inverse Compton scattering of far infrared and optical photons by high energy electrons and from relativistic bremsstrahlung of low energy relativistic electrons. The latter process is expected to be more important at low galactic latitudes.

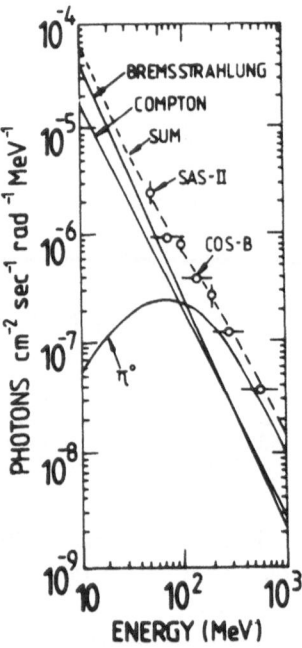

Fig. 4. The spectrum of the Galactic γ-ray emission from observations made by the SAS-II and COS-B satellites. The intensity is measured in the direction of the Galactic Centre. (Kniffen and Fichtel 1981, Ramana Murthy and Wolfendale 1986).

Relativistic bremsstrahlung is perhaps less familiar to radio astronomers than synchrotron radiation and so a few words are appropriate. Just like non-relativistic bremsstrahlung, the spectrum of relativistic bremsstrahlung is more or less flat up to an energy $h\nu \approx E$ where E is the kinetic energy of the electron; above $h\nu \approx E$, the radiation falls off very rapidly (see Longair 1990). Thus, the expected γ-ray spectrum of a power-law distribution of relativistic electrons is just the superposition of a continuous set of spectra which cut-off at energy E. A simple calculation shows that the photon spectral index α_γ is the same as the electron energy spectral index a. Typically, the energies of the γ-rays are about one-third the energy of the electron and so 30–100 MeV γ-rays correspond to roughly 100–300 MeV electrons. Thus, the low energy spectrum of the electrons is expected to be

$$N(E)dE \propto E^{-1.6}dE.$$

Does this work quantitatively? For illustrative purposes, I have carried out a simple calculation in which I assume that the γ-rays observed by SAS-2 in the 35–100 MeV waveband originate from within the inner Galaxy in a thin uniform disc similar to the HI and molecular hydrogen discs. This model shows that the electron energy spectrum with index $a = 1.6$ joins smoothly onto the electron energy spectrum inferred from the Galactic radio emission (Figure 3). This result is confirmed by much more detailed calculations carried out by Kniffen and Fichtel (1981). These calculations make it entirely plausible that the low energy γ-ray emission of the Galaxy complements the low frequency radio emission. As Dr Webber pointed out, the same electrons radiating synchrotron radiation in the Galactic magnetic field at 1 MHz radiate γ-ray bremsstrahlung at ~ 60 MeV, the first process depending on the magnetic field strength and the second on the number density of the hydrogen nuclei.

There is thus considerable interest in determining precisely both the low energy part of the γ-ray spectrum of the interstellar gas and its low frequency radio spectrum. The inferred electron energy spectral index $a = (2\alpha + 1) = 1.8$ at about 10 MHz (assuming $\alpha = 0.4$) is similar to the value 1.6 inferred from the γ-ray emission. My interpretation of the γ-ray observations is that a range of spectral indices could fit the data. The importance of observations with the Gamma Ray Observatory for this programme is obvious. Because of interstellar absorption, it becomes progressively more and more difficult to measure the radio spectrum throughout the Galaxy below 10 MHz. The spectrum can however be found by looking at the Galactic emission in the direction of an opaque nearby region of ionised hydrogen. It should also be noted that the low energy γ-ray waveband is one of the most difficult for observation.

5 Interpretation

The procedure for interpreting these observations of the Galactic electron spectrum are very well known. We need an equation which takes account of diffusion, convection, injection and losses of the electrons. The problem can be made as complicated as one wishes. To make progress let us start with the simplest version of the diffusion-loss equation for relativistic electrons

$$\frac{dN}{dt} = D\nabla^2 N + \frac{\partial}{\partial E}[b(E)N(E)] + Q(E).$$

The equation describes the evolution of the electron energy spectrum, the diffusion of the electrons being described by a scalar diffusion coefficient D and the injection spectrum by $Q(E)$. The rate of energy loss (or gain) is described by $b(E)$ and we can write this term

$$b(E) = -\frac{dE}{dt} = A_0 + A_1 E + A_2 E^2 \quad (-A_3 E).$$

These terms have the following meanings:

A_0 represents ionisation losses which depend only logarithmically upon energy and correspond to about $10N$ eV per year where N is the particle density in particles cm^{-3};

$A_1 E$ represents adiabatic and relativistic bremsstrahlung losses;

$A_2 E^2$ represents synchrotron and inverse Compton losses;

$-A_3E$ represents energy gains i.e. acceleration. This last term can represent first and second Fermi acceleration.

The very brave solve the diffusion-loss equation as it stands for a particular distribution of sources. The more timid can replace the diffusion term by a term of the form $N(E)/\tau(E)$ where $\tau(E)$ is the time scale for escape from the Galactic "confinement volume". The form of $\tau(E)$ can be chosen to mimic diffusion (if a gaussian distribution of loss times is used), or a leaky box (if an exponential distribution of loss times is used). Also, if the loss time were to depend on energy, a power-law distribution of loss times results in a simple solution for the equilibrium energy spectrum.

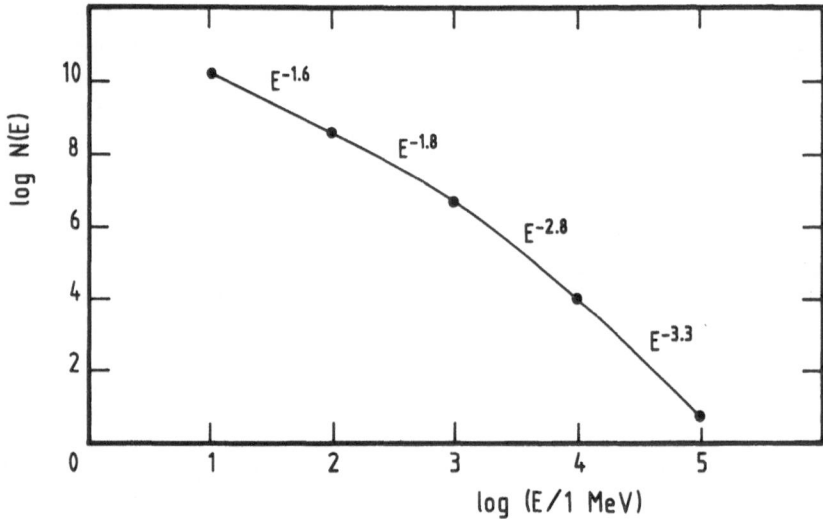

Fig. 5. A schematic representation of the electron energy spectrum in the local interstellar medium from the data discussed in this paper. This spectrum has been subject to energy losses at high and low energies during the propagation of the particles through the interstellar medium. The units of $N(E)$ are relative units (see Figure 3 for physical units).

The simplest way of interpreting the observations, which I display schematically in Figure 5, is to look first at those energy ranges in which the time scale for energy losses is less than the escape time $\tau(E)$. If the electrons have escape times $\sim (1-3) \times 10^7$ years, similar to the escape times derived for cosmic ray nuclei from the observed abundances of radioactive ^{10}Be nuclei, then the time scales for energy losses are less than this value for the lowest energy electrons (E $\lesssim$ 300 MeV) against ionisation losses and at the highest energies by synchrotron losses. For example, taking an ionisation loss rate of $10N$ eV per year, we find a lifetime of 10^7 years for 100 MeV electrons if $N = 1$ cm^{-3}; taking $B = 6 \times 10^{-10}$T, we find the lifetime of 10 GeV electrons to be about 3×10^7 years. Thus, if electrons are continuously injected uniformly into the interstellar medium, the electron spectra in these regions should reach a steady-state under losses, i.e. $dN/dt = 0$ and we have to solve

$$\frac{\partial}{\partial E}[b(E)N(E)] = -Q(E)$$

If $Q(E) = E^{-a}$,

$$N(E) = \frac{KE^{-(a-1)}}{(a-1)b(E)}.$$

Thus, in the low energy regions in which ionisation losses dominate $N(E) \propto E^{-(a-1)}$ and so the injection spectrum would be $Q(E) \propto E^{-2.6}$. In the high energy region, in which synchrotron losses dominate, $N(E) \propto E^{-(a+1)}$ and hence the injection spectrum is $Q(E) \propto E^{-2.3}$. These values are not too different and this suggests that the injection electron spectrum might be quite close to $E^{-2.5}$ throughout the energy range 100 MeV to 100 GeV.

Breaks in the spectra are expected at the energies at which the lifetimes of the electrons are equal to their escape times. Thus, the break in the radio spectrum at 1-2 GHz (see Figure 2) can be associated with the equality of the lifetimes of the electrons radiating at these energies and the escape time τ. Likewise, for the low energy electrons, the break would reach to $E \approx 300$ MeV in 3×10^7 years if $N \sim 1$ cm^{-3} corresponding to $\nu = 2$ MHz if $B = 6 \times 10^{-10}$ T. There are many reasons why the spectra of the electrons and the radio spectra would be smoothed out over a wide range of energies and frequencies respectively about these critical values. These calculations are schematic but I make them to illustrate the point that only in the intermediate range of energies at $E \sim 1$ GeV, should the spectrum be "stable".

At energies at which the loss time is of the same order as the synchrotron loss time, we might begin to see variations in the radio spectral index of the Galaxy in different directions. This is a notoriously difficult observation because whole sky surveys have been made with telescopes of very different resolution and polar diagram. Lawson et al (1987) have made an attempt to do this for whole sky radio surveys in the frequency range 150 MHz to 5 GHz. They find little evidence for spectral variations at low radio frequencies but find evidence for significant fluctuations at high frequencies. This would be consistent with the idea that significant variations occur at frequencies at which the loss time is of the same order as or less than the escape time. By the same logic, there may well be significant fluctuations in the very low frequency spectrum of the Galaxy when the ionisation losses dominate. Incidentally, Banday and Wolfendale (1990) point out that these variations in the high frequency spectral index of the Galaxy can confuse searches for fluctuations in the Microwave Background Radiation at high sensitivities.

6 Conclusions

There are plainly questions of real astrophysical importance which can be addressed by very low frequency radio surveys. The paramount considerations are high angular resolution and high sensitivity. In general, a dominant theme is the determination of the spatial distribution of relativistic electrons. The unique feature is the availability of well defined path lengths over which the emissivity can be determined. To put it another way, these observations would provide 3-dimensional measures of the distribution of high energy electrons and magnetic fields rather than 2-dimensional projections onto the sky.

From this basic set of observations, a number of studies follow immediately: the lifetimes of electrons in the interstellar gas and the competition between escape and energy loss; evidence on electron acceleration processes; the ratio of high energy electrons to protons in the interstellar medium and in their sources. An exciting new field will be to combine these observations with low energy γ-ray studies which should provide complementary evidence on the same relativistic electrons.

Acknowledgements

I thank Professor A.W. Wolfendale and Dr X. Chi for helpful comments on the first version of this paper.

References

1. Baldwin, J.E., 1967. In "Radio Astronomy and the Galactic System", ed. H. van Woerden, p. 337, IAU Symposium No. 31, Academic Press.
2. Banday, A.J. and Wolfendale, A.W., 1990. Mon. Not. Roy. astr. Soc., (in press).
3. Bridle, A.H., 1967. Mon. Not. Roy. astr. Soc., 136, 219.
4. Bridle, A.H., 1968. Mon. Not. Roy. astr. Soc., 138, 251.
5. Caswell, J.L., 1976. Mon. Not. Roy. astr. Soc., 177, 601.
6. Kassim, N., 1990. see this volume.
7. Chi, X. and Wolfendale, A.W., 1990. J. Phys. G. (submitted).
8. Kniffen, D.A. and Fichtel, C.E., 1981. Astrophys. J., 250, 389.
9. Lawson, K.D., Mayer, C.J., Osborne, J.L. and Parkinson, M.L., 1987. Mon. Not. Roy. astr. Soc., 225, 307.
10. Longair, M.S., 1990. "High Energy Astrophysics", second edition, Cambridge University Press.
11. Purton, C.R., 1966. Mon. Not. Roy. astr. Soc., 133, 463.
12. Ramana Murthy, P. and Wolfendale, A.W., 1986. "Gamma Ray Astronomy", Cambridge University Press.
13. Rand, R.J. and Kulkarni, S.R., 1989. Astrophys. J., 348, 760.
14. Reynolds, R., 1990. see this volume.
15. Stecker, F.W., 1976. In "The Structure and Content of the Galaxy and Galactic gamma-rays, ed. C.E. Fichtel, Goddard Space Flight Center Publications.
16. Webber, W., 1983. In "Composition and Origin of Cosmic Rays", ed. M.M. Shapiro, 83, D. Reidel Publ. Co.
17. Webster, A.S., 1971. PhD dissertation, University of Cambridge.

VIII. EXTRAGALACTIC LOW FREQUENCY ASTROPHYSICS

KNOWN AND EXPECTED SOURCES OF LOW-FREQUENCY RADIATION

Wolfgang Kundt

Institut für Astrophysik der Universität

Auf dem Hügel 71, D-5300 Bonn 1, FRG

Abstract: A prediction of the astrophysical sources at low radio frequencies requires a deep understanding of the cosmic engines. Even when coherent processes are ignored, there are a number of uncertainties due to alternative explanations of the known sources, such as supernovae and supernova shells, the bipolar-flow family, and the galactic chimneys. An attempt is made to classify them as either magnetic centrifuges or magnetic reconnectors. It is not clear whether or not we have identified the cosmic-ray boosters; low-frequency observations may help disclose them.

1. Constraints on Observability

In order to observe an astrophysical source at some frequency ν, this frequency must be above the resonances of the intervening media, and the media must not be optically too thick at ν. For the warm component of the ISM, these two conditions read

$$\nu > \nu_p + \nu_B/2 \approx 10 \text{ KHz } n_0^{1/2} \tag{1}$$

and

$$\tau(\nu) = (\int n_e^2 ds)_{18.8} / T_4^{3/2} \nu_6^2 \lesssim 1 . \tag{2}$$

Here $\nu_p = (n e^2/\pi m_e)^{1/2}$ is the (electron) plasma frequency, $\nu_B = eB/2\pi m_e c$ the (electron) gyro frequency at field strength $B(\lesssim 10^{-5} \text{ G})$, $n_0 := n_e/10^0 \text{ cm}^{-3}$, $\int n_e^2 ds =: \text{EM}$ is the emission measure, $(\text{EM})_{18.8} := \text{EM}/10^{18.8} \text{ cm}^{-5}$, and $\nu_6 := \nu/\text{MHz}$. The emission measure of the ISM is of order 10^{19} cm^{-5}, i.e. the ISM is transparent down to $\nu \approx \text{MHz}$. But HII-regions will show up in absorption, with optical depth τ (MHz) = $10^{2.5 \pm 1}$.

Astrophysical sources can be of thermal or non-thermal origin. The former are intrinsically bright at low radio frequencies only if large in extent, larger than ordinary stars. This communication will be restricted to non-thermal sources. Among them, coherent sources - like pulsars - are difficult to predict.

For incoherent non-thermal sources, the literature tends to ignore the influence of the medium in which the source is embedded. The standard expressions for both synchrotron and Čerenkov radiation are not applicable when the refractive index n ($=\beta_{wave}^{-1}$) differs as much (or more) from unity as the radiating particle's $\beta := v/c$. In such cases, the relevant dimensionless variable x occurring in the emission integrals reads [Schwinger et al, 1976]:

$$x := \left[(1-n^2\beta^2)/(1-\beta^2)\right]^{3/2} \nu/\gamma^2 \nu_{\mathrm{B}} \approx \left[1+(\gamma\nu_p/\nu)^2\right]^{3/2} \nu/\gamma^2 \nu_{\mathrm{B}} , \qquad (3)$$

the latter for $\gamma \gg 1$,, $\nu \gg \nu_p$, ν_{B} (because of $n^2 \approx 1-(\nu_p/\nu)^2$) . In vacuum, we have $x = \nu/\gamma^2 \nu_{\mathrm{B}}$, which leads to the well-known upper synchrotron cutoff at $\nu = \gamma^2 \nu_{\mathrm{B}}$ (for x = 1). In a plasma with $\nu \gg \nu_{\mathrm{B}}$, x cannot decrease below unity - i.e. radiation is exponentially suppressed - unless

$$\gamma > \gamma_{\mathrm{R}} := \nu_p/\nu_{\mathrm{B}} = n_0^{1/2}/B_{-2.7} , \qquad (4)$$

where γ_{R} is the Razin-Lorentz factor. Once γ is larger than γ_{R}, a low-frequency cutoff results (again) from the condition $x \leq 1$, viz $\nu \geq \nu_{\mathrm{R}}$:= Razin frequency:

$$\nu_{\mathrm{R}} = (\gamma\nu_p^3/\nu_{\mathrm{B}})^{1/2} = 0.5 \text{ GHz } (\gamma_4 \, n_4^{3/2} / B_{-2})^{1/2} . \qquad (5)$$

In other words: synchrotron radiation is exponentially suppressed for Lorentz factors below γ_{R} and for frequencies below ν_{R}, see fig. 1. This condition has been proposed to explain the steep turnover in quasar spectra, at wavelengths between 0.1 and 1 mm [Schlickeiser & Crusius, 1989]. It may equally apply to Sgr A West in our Galactic nucleus, for the parameter values inserted into eq. (5) [Kundt,1990b]; and it may be of relevance to the filaments of supernova shells. Low-frequency cutoffs can therefore give us information on the properties of the emission region.

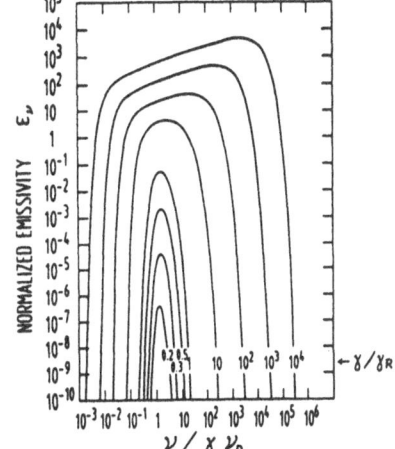

Figure 1: Synchrotron emissivity ε_ν (in arbitrary units) versus frequency ν inside a weakly magnetized plasma , taken from Crusius & Schlickeiser (1988). The Razin-Lorentz factor γ_R and plasma frequency ν_p are given in the text. Note that ε_ν vanishes exponentially for $\gamma < \gamma_R$, and for $\nu < \nu_R$ when $\gamma \gtrsim \gamma_R$.

2. Boosters for Relativistic Electrons

Three types of engine have been considered in the literature as the possible accelerators of electrons - and perhaps positrons - to relativistic energies, i.e. as the engines of the non-thermal sources, cf. fig. 2. All of them involve time-variable magnetic fields, listed below with a decreasing amount of large-scale order. They are

(1) The Magnetic Centrifuge: A rotating magnet, or magnetized rotator accelerates charges of either sign near and beyond its speed-of-light cylinder. This booster has been introduced by Gunn and Ostriker to explain pulsars [Kulsrud et al, 1972; Camenzind, 1990]. Doubts have been occasionally expressed in its efficiency, because of plasma quenching. Such doubts are unlikely to apply to situations where magnetic pressures exceed plasma pressures inside the corotating magnetosphere, simply because of radial-momentum conservation: The changing electromagnetic fields must share their radial momentum with the threaded charges. In my opinion, the Gunn-Ostriker mechanism is at work not only in the (radio-) pulsars but also in all sources of the bipolar-flow family [Kundt, 1987a].

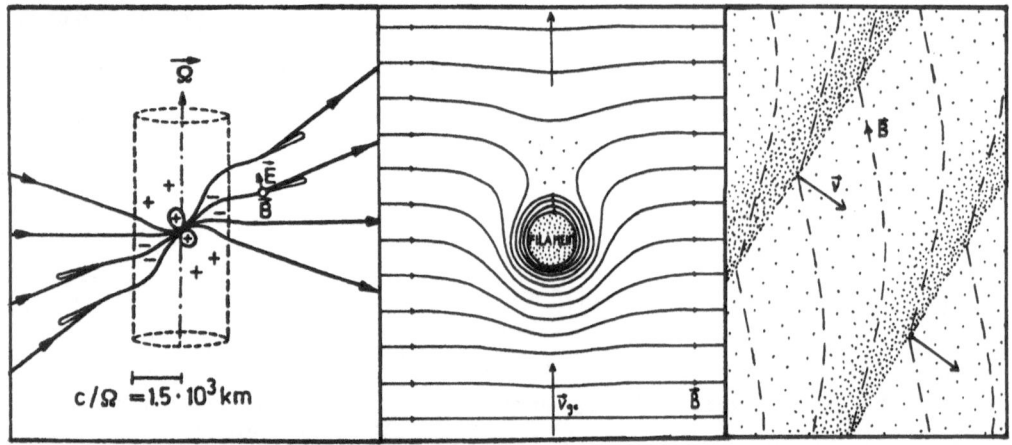

Figure 2: The three types of proposed cosmic particle accelerators: (1) magnetic centrifuge (Crab PSR); (2) magnetic reconnection (behind a filament); and (3) diffusive
shock acceleration.

(2) Magnetic Reconnection: Magnetic fields are frozen into highly
conducting plasmas and strained until their pressures reach the ram
pressures of the medium. Whenever low densities and/or poor
conductivities permit transition towards a force-free configuration,
magnetic fields recombine and transfer their energy to the medium.
Important examples are solar flares, solar-wind corotating interaction
regions, bowshocks and magnetotails as well as high-density SN
filaments crossing magnetized winds, and perhaps supernovae [Kundt,
1990a].

(3) Diffusive Shock Acceleration: This mechanism has been widely
explored for more than ten years, partly because there are
astrophysical sources whose relativistic charges have deceleration
lifetimes much shorter than transit times and partly because Earth is
continually bombarded by (cosmic-ray) particles with energies up to
and in excess of 10^{20} eV, whose boosters have so far evaded detection.
One then invokes in-situ, or Fermi acceleration for a replenishment of
the energy losses. Diffusive shock acceleration is a mechanism to
segregate and prevent a small sub-population of particles from joining

the thermalization draught, but its efficiency may have been largely overestimated [Kundt (1984), Falle (1990)]. In many cases, in-situ acceleration can be replaced by in-situ deceleration of particles from an otherwise loss-free stream which collide with small obstacles. The compact central engines tend to be vastly more efficient than the extended emission sites.

3. Supernovae and Supernova Shells

In the literature, supernovae of type II are often thought to be driven by the neutrinos liberated (thermally) during core collapse of a massive star. There is, however, a minority view that instead, supernovae are driven by a transfer of the collapsing core's angular momentum via magnetic tensions [Hoyle (1946), Bisnovatyi-Kogan et al (1976), Kundt (1990a)]. The neutrinos are unlikely to transfer enough radial momentum, and supernova shells have the filamentary morphology reminiscent of magnetic Rayleigh-Taylor instabilities.

If supernova explosions are driven by magnetic shear forces, there are two important consequences [Kundt, 1990a]: 1) Reconnection of the huge magnetic fluxes will give rise to ample production of relativistic pair plasma; and 2) the ejected shell will be squeezed and torn, in the supernova event itself, into thousands of bits and pieces without pressure contact. Supernova explosions would thus resemble (thick-walled) splinter bombs rather than pressure bombs. Both consequences lend themselves to multiple tests by the observations. For instance, filaments of SN shells may already have made their appearance via scintillation events [Fiedler et al (1987), Wolszczan & Cordes (1987)] and/or via the IR cirrus seen at high galactic latitudes.

Among others, a supernova should be detectable at radio frequencies as soon as its progenitor star's windzone gets transparent. The condition reads:

$$1 \geq \tau(\nu) = (EM)_{26.5}/T_5^{3/2} \nu_9^2 = \begin{cases} 10^{-2.5} (\dot{M}_{(-6)}/v_8)^2 \\ \\ 10^{3.5} (\dot{M}_{(-5)}/v_6)^2 \end{cases} r_{15}^{-3} \nu_9^{-2} , \qquad (6)$$

where the first line is characteristic of a blue (super-) giant's wind, the second line of a red giant's wind, both at a radius of r = 10^{15} cm corresponding to an age of $\gtrsim$ 10 days; $\dot{M}_{(-6)} := \dot{M}/10^{-6} M_\odot$ yr^{-1}. Clearly, red-giant windzones take much longer to become transparent than blue-giant ones. There is no apriori need for a supernova to stay on for long at radio frequencies, once the shell gets transparent, because a cloud of relativistic electrons (and positrons) can expand adiabaticly with small radiation losses when there are no obstacles.

Some or (almost) all supernovae leave neutron stars behind - which we see as pulsars if isolated and strong enough, but hardly if screened by the windzone of a near companion. The binary (or multiple) ones become X-ray sources after sufficient spindown (t $\gtrsim$ 10^4 yr). Pulsars have very soft radio spectra. Their low-frequency cutoffs have not been found in all cases, most notably for the Crab pulsar, fig. 3. It is important to know whether there is a strict correlation between a pulsar's low-frequency and high-frequency turnover, as proposed by Malofeev & Malov (1980).

Supernova shells are visible for $\lesssim$ 10^4 years. Older shells are difficult to map, as show the examples of CTB 80 [Strom (1988), Fesen et al (1988), Kundt (1990a)] and W 30 [Kassim & Weiler, 1990]. Even the (10^4 yr old) Vela SNS is often mapped incompletely: its pulsar should sit right at the center, which it apparently does when the whole shell is mapped, whose brightness is far from uniform [Kundt, 1990a]. Old shells tend to show up more distinctly at IR- and, probably, at low radio frequencies.

A young shell whose true size is open to doubt is the Crab nebula. Fig. 4 shows schematicly its different components, traced by thermal continuum, non-thermal, and emission-line radiation. If the visible

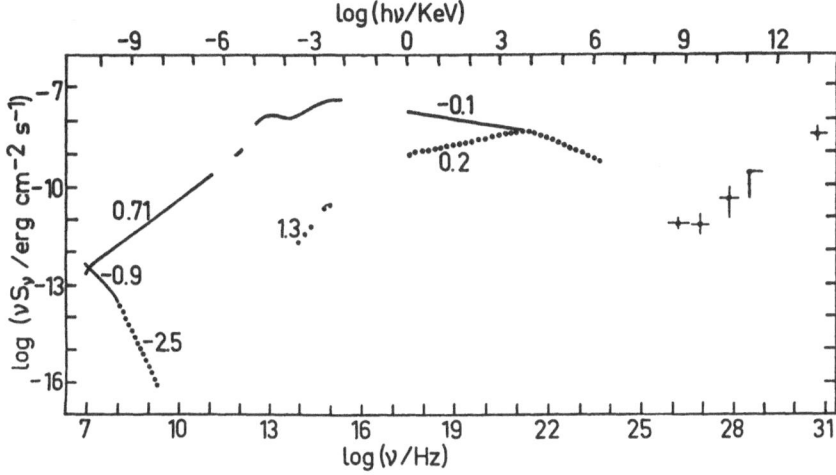

Figure 3: Spectrum of the Crab pulsar (dotted) and its nebula, νS_ν versus ν. At the low-frequency end , the pulsar is smeared out by dispersion.

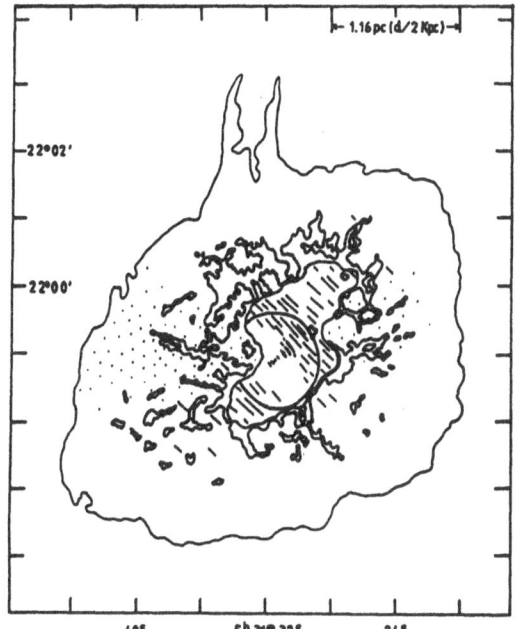

Figure 4: Schematic drawing of the Crab nebula and its pulsar, as seen on various line- and continuum-maps;taken from Kundt (1990a).

outer edge were the true outer edge of its ejecta, the Crab would be deficient in kinetic energy by a factor of $\gtrsim 30$ (compared with the 10^{51} erg of a typical supernova). I have argued in [1990a] that the Crab's progenitor has likely been a blue supergiant with a low-density wind, and that we have not mapped the outer two thirds (in radius) of its ejecta: they are of too low density to be visible in either emission or absorption. No relativistic electrons have escaped from the mapped shell, except through the 'jet'; hence there is no hope - it is claimed - of ever mapping the outer two thirds at radio frequencies.

4. The Bipolar-Flow Family

Four classes of astrophysical objects show similar morphologies, high velocities, and broad spectra: 1) the extragalactic radio sources, apparently powered by active galactic nuclei (=AGN); 2) the young stellar objects (YSO): IR-detectable pre-T-Tauri stars in star-forming regions which power the 'bipolar flows'; 3) certain binary neutron stars, such as Sco X-1, Cyg X-3, SS 433, and others; and 4) certain binary white dwarfs, such as CH Cyg and R Aqu, with or without surrounding planetary nebulae. Fig. 5 shows one representative of each.

Common properties of the bipolar-flow family are: (i) elongated morphologies, inferred to range through distance ratios of $10^{6\pm2}$ and more, containing narrow, knotty jets near their long axis; (ii) highly supersonic speeds, with core velocities exceeding several percent of the speed of light, even in the 'butterfly nebula' M 2-9 [Balick, 1989; see also 1987] (though not (yet) in YSOs); (iii) broad spectra, from low radio frequencies through IR-, optical and UV-frequencies up to (often) X-ray energies, γ-ray energies and even UHE γ-ray energies; and (iv) core/lobe power ratios in the range $10^{2\pm2}$, implying average efficiencies in jet formation of order 1% and strongly variable central engines.

In the literature, very different explanations have been given for individual sources of the BF family. There is a minority claim that all of them contain a magnetic centrifuge which creates relativistic $e^{\pm}$ pairs in localized magnetospheric discharges: Kundt & Gopal-Krishna (1984) for Sco X-1, Kundt (1987b) for SS 433, Blome & Kundt (1988) for the YSOs, Kundt & Fischer for the black-hole candidates, and Kundt

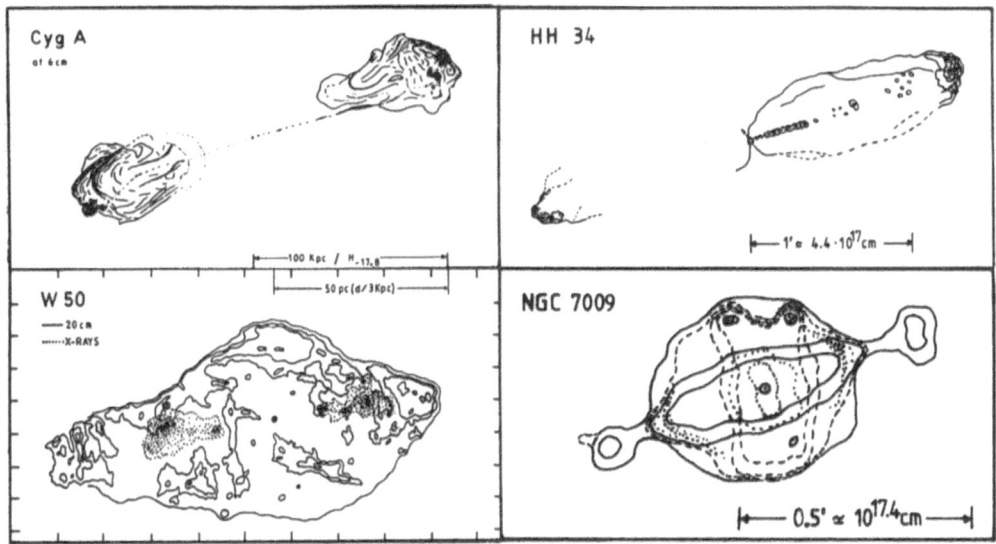

Figure 5: Representatives of the Bipolar-Flow family: (1) radio negative (drawing) of Cyg A , the brightest radio galaxy; (2) optical CCD contour map (drawn) of HH 34 , a young bipolar flow; (3) radio isophotes and X-ray plot (stippled) of the old SNS W50 and the bipolar outflow from the neutron-star binary SS 433; (4) multicolour optical drawing of the Saturn nebula, NGC 7009, in Aquarius, a middle-aged planetary nebula; after Balick (1987).

(1987a) for the whole class. The jets of Cyg X-3 have been recently mapped by Strom et al (1989). Radio emission of highly non-thermal brightness has been found for 3 T-Tauri stars by Feigelson et al (1990). The recently reported subluminal velocities in the jet of M 87 [Reid et al (1989), Owen et al (1989)] need only be phase velocities: bulk Lorentz factors of order $\gamma \gtrsim 10^4$ are indicated for independent reasons [Kundt, 1989a]. The jet of M 87 is presently the best example of highly non-isotropic radiation from a thin boundary layer of the beam.

If all the bipolar-flow sources create relativistic pair plasma, their aged emission lobes should be strong emitters at low radio frequencies. In the case of our Galactic center, this expectation has been verified by La Rosa & Kassim (1985); cf. Kundt (1990b).

5. The Cosmic Rays

The origin of the cosmic rays has been a problem ever since their discovery, in particular for high particle energies (in excess of some 10^{15} eV). Mildly or moderately relativistic particles are produced at the Earth's bowshock, or in solar flares. But particle energies above 10^{12} eV are difficult to obtain even by much more violent stellar winds, or by supernova ejecta; occasional claims to the contrary have not been convincing[Kundt, 1989b].

It is normally argued that only a small percentage of the cosmic-ray energy is contained in the particles of very high energy. Figure 6 shows that this is true for the observed cosmic-ray spectrum. It is not true, however, for the source spectrum, i.e. for the spectrum to be generated by the cosmic-ray boosters.

In order to see this, note that typical cosmic-ray particles have a residence time of $10^{7 \pm 0.5}$ years in the gaseous disk of the Galaxy before they escape through the halo into intergalactic space. This lifetime may be energy-dependent, in proportion to E^{-x} where x is estimated between 0 and 0.6 for low energies, depending on one's (leaky box) model. For energies of 10^{16} eV, say, the average escape time from the disk should still exceed 10^2 gyrations (of radius $R_B = 1$ pc γ_7 each), corresponding to a minimum residence time of $10^{3.5}$ years; this limits the source function. But at energies beyond 10^{19} eV, the escape time should shrink to the transit time, of order $10^{2.5}$ years, implying a $10^{4.5}$ times more powerful injection (than at low energies) if the boosters are galactic.

At these highest cosmic-ray energies, researchers often prefer an extragalactic origin; but then the situation gets worse: The storage volume is now the whole of intergalactic space, whose volume exceeds that of galactic disks by their inverse mass density ratio $\rho_{disk}/\rho_{cosmic} = 10^{7 \pm 1}$. At particle energies above 10^{20} eV, the collisional lifetime is limited (by the background radiation) to 10^7 years, hence the injection must exceed that at low energies by a factor of 10^7. I thus arrive at the dotted injection spectrum of fig. 6. Note that the relative source power at energies $> 10^{19}$ eV amounts to $10^{-2 \pm 1}$, an embarrassingly large fraction.

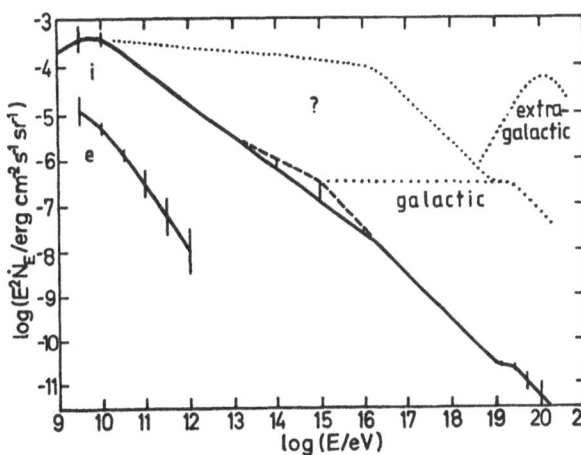

Figure 6: Energy distribution of the cosmic rays, $E^2 \dot{N}_E$ versus E, for both electrons and ions, together with inferred source spectra of the ions (dotted). Larger efficiencies (powers) would be required if the ultrahigh-energy sources were extragalactic.

Aspects of this problem have been recently discussed by Berezinsky & Grigor'eva (1988) and by Stecker (1989). On the observational side, the long-standing claim of structured anisotropies at ultrahigh energies (E $\gtrsim 10^{17}$ eV) appears to have been eliminated by Fly's Eye [Cassiday et al, 1989]. We therefore search for powerful boosters, isotropically arranged around the solar system, which cannot be supernovae. What are they? I should not be surprised if the answer read: clumpy accretion by neutron stars in the halo [Kundt et al, 1987]. Shouldn't such sources show up at very low radio frequencies?

6. The Galactic Chimneys

A recent discovery in both our and neighbouring galaxies are "worms", "filaments", "threads", or "spurs" sticking out by hundreds of parsecs perpendicular to the disk. They have become popularized under the name of "chimneys" by Norman & Ikeuchi (1989) and explained in terms of outflows powered by supernovae and/or stellar winds.

Galaxies are thought to contain hundreds of them, seen in HI emission, dust absorption, Balmer emission, and radio emission. A good example is shown in fig. 7.

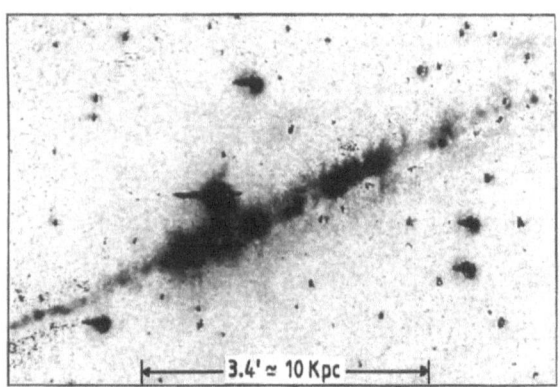

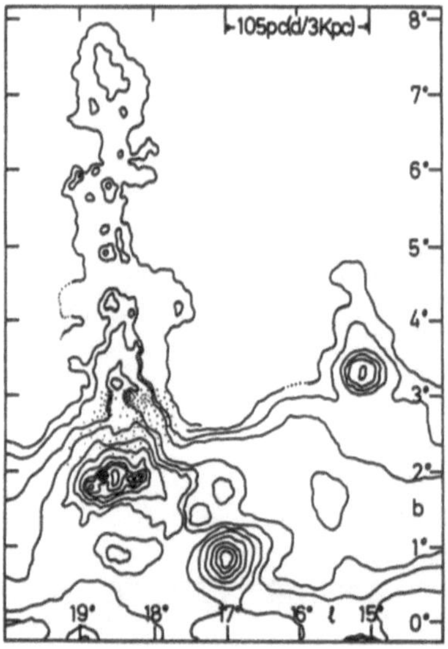

Figure 7: Hα CCD negative of the galaxy NGC 891, showing some twenty chimneys in emission ;   from Dettmar et al (1990).

Figure 8: Stockert's chimney , the nearest iden- tified galactic chimney, mapped at 2.7 GHz (isophotes) and CO emission (stippled).

The nearest identified such worm is "Stockert's chimney", a thermal radio column at a distance of (2.9±0.4) Kpc [Kundt & Müller, 1987], see fig. 8. Obscured by foreground dust, its optical appearance is not known. Stockert's chimney is powered by the bright H II-region S 54 and is fed through a molecular nozzle of diameter 25 pc, to a height of 300 pc above the disk. If made of stellar material, it would plainly sink into the disk under its own weight. No recoil is visible, hence it cannot have been thrown up either.

The conclusion, therefore, is that the galactic chimneys are supersonic jets driven by a (nearly) weightless substance, most likely by pair plasma (from the bipolar flows inside of S 54) plus cosmic rays plus some hot gas which is dragged along like sand by a storm. They form the leak for the cosmic rays and blow the galactic halo. They turn the galactic halo into a non-thermal radio source.

Acknowledgements: Conversations with Reinhold Schaaf, Peter Biermann, Daniel Fischer, Axel Jessner, and Gerrit Verschuur are gratefully acknowledged.

References

Balick, B., 1987: Sky & Telescope, February, 125
Balick, B., 1989: Astron. J. 97, 476
Berezinsky, V.S. & Grigor'eva, S.I., 1988: Astron.Astrophys. 199, 1
Bisnovatyi-Kogan, G.S., Popov, Yu.P. & Samochin, A.A., 1976: Astrophys.Space Sci. 41, 287
Blome, H.-J. & Kundt, W., 1988: Astrophys.Space Sci. 148, 343
Camenzind, M., 1990, in: NATO ASI C 300, ed. W. Kundt, Kluwer, p. 139
Cassiday, G.L. et al (16 authors), 1989: UUHEP-89/4, Salt Lake City
Crusius, A. & Schlickeiser, R., 1988: Astron.Astrophys. 196, 327
Dettmar, R.-J., Keppel, J., Roberts, M. & Gallagher, J.S. III, 1990, in: The ISM in external galaxies, 2nd Wyoming Conf., eds. D. Hollenbach & H. Throtnsen, NASA
Falle, S.A.E.G., 1990, in: NATO ASI C 300, ed. W. Kundt, p. 303
Feigelson, E.D., Lonsdale, C.L. & Phillips, R.B., 1990: AAS contribution 20.07
Fesen, R.A., Shull, J.M. & Saken, J.M., 1988: Nature 334, 229
Fiedler, R.L., Dennison, B., Johnston, K.J. & Hewish, A., 1987: Nature 326, 675
Hoyle, F., 1946: Mon.Not.R.astr.Soc. 106, 343
Kassim, N.E. & Weiler, K.W., 1990: Nature 343, 146
Kulsrud, R.M., Ostriker, J.P. & Gunn, J.E., 1972: Phys.Rev.Lett. 28, 636
Kundt, W., 1984: J.Astrophys.Astron. 5, 277
Kundt, W., 1987a: Astrophysical Jets and their Engines, NATO ASI C 208, Reidel, p. 1
Kundt, W., 1987b: Astrophys.Space Sci. 134, 407
Kundt, W., 1989a, in: Hot Spots in Extragalactic Radio Sources, Lecture Notes in Physics 327, eds. Meisenheimer & Röser, Springer, pp. 179, 275
Kundt, W., 1989b: Adv.Space Res. 9, (12) 81
Kundt, W., 1990a: Neutron Stars and their Birth Events, NATO ASI C 300, Kluwer, p. 1
Kundt, W., 1990b: Astrophys.Space Science, submitted
Kundt, W. & Fischer, D., 1989: J.Astrophys.Astron. 10, 119
Kundt, W. & Gopal-Krishna, 1984: Astron.Astrophys. 136, 167
Kundt, W. & Müller, P., 1987: Astrophys.Space Sci. 136, 281
Kundt, W., Özel, M. & Ercan, E.N., 1987: Astron.Astrophys. 117, 169
La Rosa, T.N.& Kassim, N.E., 1985: Astrophys.J. 299, L 13
Malofeev, V.M. & Malov, I.F., 1980: Sov.Astron.24 (1), 54.
Norman, C.A. & Ikeuchi, S., 1989: Astrophys.J. 345, 372
Owen, F.N., Hardee, P.E. & Cornwell, T.J., 1989: Astrophys.J. 340, 698
Reid, M.J., Biretta, J.A., Junor, W., Muxlow, T.W.B. & Spencer, R.E.: 1989, Astrophys.J. 336, 112
Schlickeiser, R. & Crusius, A., 1989: IEEE Transactions on Plasma Sci. 17, 245
Schwinger, J., Tsai, W.-Y. & Erber, T., 1976: Ann.Phys. 96, 303
Stecker, F.W., 1989: Nature 342, 401
Strom, R.G., 1988, in: Supernova Shells and their Birth Events, Lecture Notes in Physics 316, ed. W. Kundt, Springer 1988, p. 91
Strom, R.G., van Paradijs, J. & van der Klis, M., 1989: Nature 337,234
Wolszczan, A. & Cordes, J.M., 1987: Astrophys.J. 320, L 35

EXTRAPOLATING ELECTRON SPECTRA TO LOW ENERGIES

D. E. HARRIS

Harvard-Smithsonian Center for Astrophysics
60 Garden St, Cambridge, MA 02138

1.0 BASICS

The Basic Question

Is "extragalactic radio astronomy below 10 MHz" a contradiction in terms? Do electron spectra follow "observed" power laws to Lorentz energy factors (γ) less than 1000, i.e. down to 100? If most non-thermal radio sources have magnetic field strengths $\leq 10^{-4}$G, then we mostly study electrons with $\gamma > 1000$. Since observations made in space will be capable of observing at frequencies more than ten times lower than those available from the ground (fig. 1), we will discuss the electron spectrum between $\gamma = 100$ and 1000.

The Basic Problem

If you told a physicist that radio astronomers measure the continuum synchrotron spectrum of non-thermal sources, he might expect that most of our papers would contain plots of the energy distribution of relativistic electrons. This is not the case because the magnetic field strength is not well determined, and because we are never quite sure that we are examining the same emission volume as we change frequency.

With the advent of aperture synthesis techniques, we have come to expect high quality maps with good resolution and high dynamic range. As we move to frequencies below 300 MHz, however, we are walking into a thick fog. Not only is high resolution

EMISSION FREQUENCY

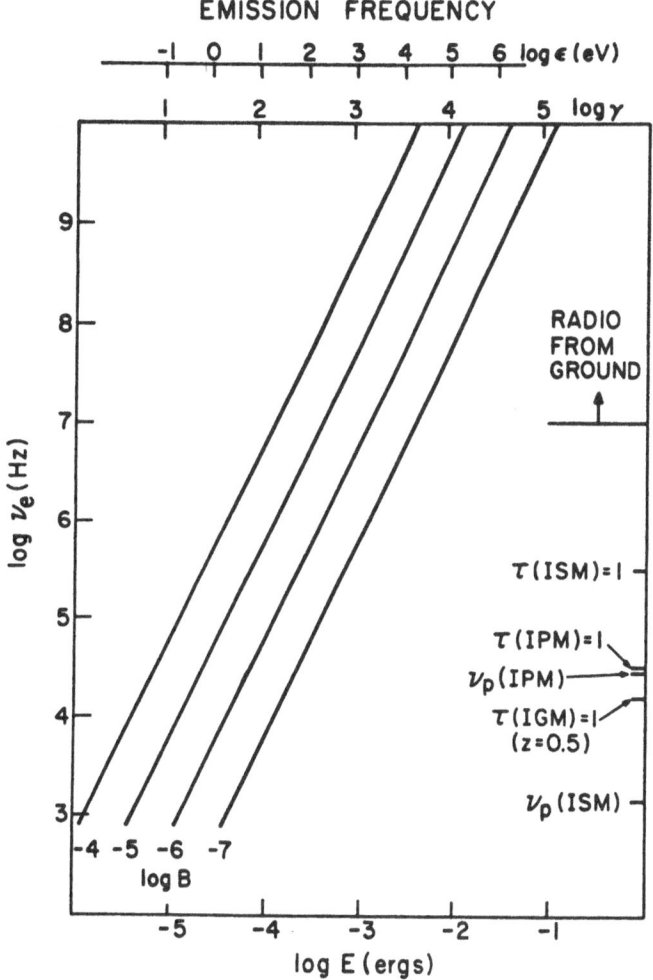

Figure 1. Emission Frequencies for Relativistic Electrons. The vertical scale shows the emitted synchrotron frequency as a function of electron energy and magnetic field strength. The top scale shows the observed frequency (energy) of IC/3K photons. Plasma frequencies and optical depths (lower right) are discussed in Harris 1987.

harder and harder to achieve, but the flux density scale degenerates to order of magnitude estimates. Brave souls occasionally build instruments with capabilities below 100 MHz, but often the results are not repeatable because the instrument is unique or it is abandoned prematurely. Over ten years ago, Braude published spectra between 10 and 25 MHz (Braude et al. 1979). Some 5-10% of these appear to have a C+ type spectrum, with the spectral index, α, greater than unity below 25 MHz (flux density,

$S \propto \nu^{-\alpha}$). To the best of our knowledge, these results have never been verified with other instruments. Between 25 and 75 MHz, we currently have only limited methods of checking our results from the Clark Lake Radio Observatory "TPT" which operated in that frequency range for a number of years.

2.0 ARGUMENTS FOR A DEPARTURE FROM THE OBSERVED POWER LAWS

In this section, we examine some of the reasons why electron spectra may deviate from the power laws inferred from the observed radio spectra. We will not be concerned with absorption which affects the observed radio spectrum; see Baldwin (this volume), for a discussion of synchrotron self-absorption.

2.1 Deviations from Observed Power Laws

Carilli et al. (1990) have made a detailed study of spectra throughout Cyg A. For two of the hot spots, they are able to fit a classical model of injection plus synchrotron losses with the break frequency between 5 and 10 GHz: α (low) = 0.5 and α (high) =1.0. However, this succeeds only for fits above 500 MHz: the 327 MHz (VLA) and 151 MHz (MERLIN) data are below the fit spectrum. Carilli believes that this is evidence for a cutoff in the electron spectrum for γ in the range 700 to 800, a model which provides a satisfactory fit to the data for a field strength equal to the minimum energy field of 300μG.

Attempts to explain the observed deviations from a power law by invoking synchrotron-self-absorption (SSA) with an angular diameter = 0.1″ and a turnover frequency close to 200 MHz, lead to very large values of the magnetic field strength: B(SSA)=13 G. Allowing the angular diameter to decrease will increase the internal pressure and magnetic field strength. If we force a SSA model in this way, a filling factor of approximately 0.001 would be required, giving a characteristic diameter of 0.01 arcsec and a field strength of 2000μG.

ELECTRON ENERGY LOSSES

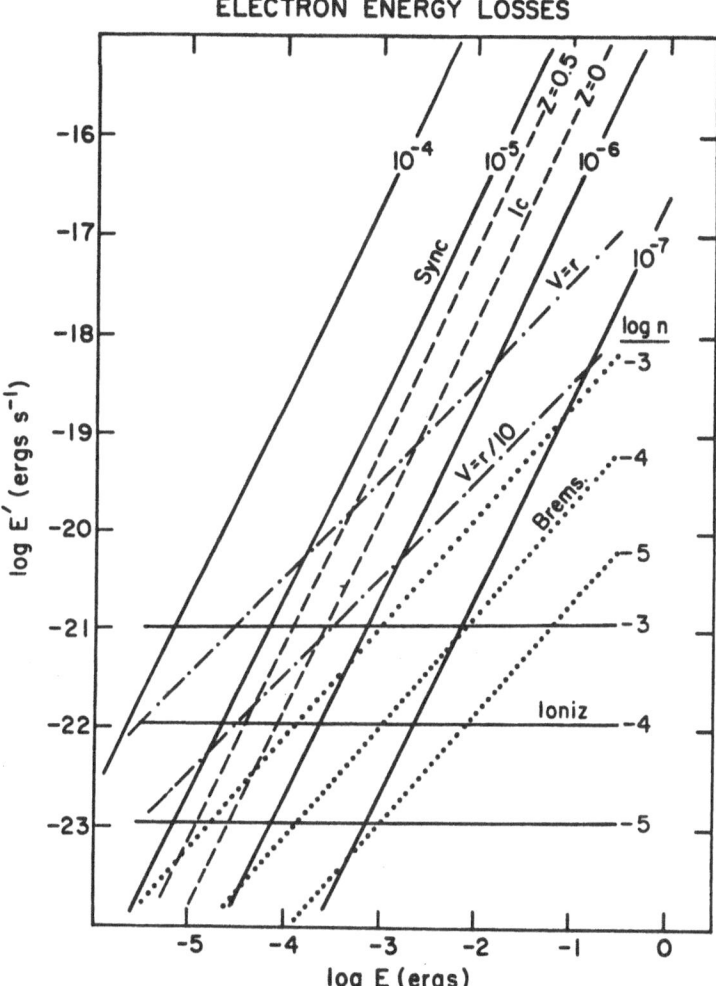

Figure 2. Energy Losses. The lines of slope 2 are for synchrotron and IC/3K losses, the former labelled with the magnetic field strength and the latter with the redshift. Dash/dot lines are for expansion losses, labelled with V, velocity in km/s, and r, radius in kpc. Values of log n, the ambient density in particles cm^{-3} label dotted lines for bremsstrahlung losses and solid lines for ionization losses.

2.2 Ionization Losses

If we examine the equations for ionization and bremsstrahlung losses (Pacholczyk, 1970), we see that bremsstrahlung losses depend on the first power of the electron energy and hence have no energy dependent effect on the spectral distribution of a power law.

$$\mathrm{dE/dt(brems)} \ = \ 1.64 \times 10^{-16} n\, E\, (0.36 + ln\gamma) \qquad \mathrm{erg/s}$$

$$\mathrm{dE/dt(ioniz)} \ = \ 1.22 \times 10^{-20} n(73.4 + ln\gamma - ln\, n) \quad \mathrm{erg/s}$$

where n = ambient density (cm^{-3}). However, the ionization losses have only a $ln E$ term, and thus the lower energy electrons have a shorter halflife than the higher energy electrons. Since both losses contain the ambient density, we see that the transition from bremsstrahlung to ionization dominance occurs at $\gamma = 1000$ $(E = 8.2 \times 10^{-4}$ erg$)$. As is evident from Figure 2, this transition energy has only a weak dependence on the ambient density.

The depletion of low energy electrons will be evident only in sources for which the ionization losses dominate synchrotron, inverse Compton, and other losses. This may occur for old sources in weak field regions: e.g. $B < 3 \times 10^{-6}$ Gauss, and $n > 10^{-3}\ \mathrm{cm}^{-3}$ (Figure 3).

2.3 E^2 Losses

If synchrotron and/or inverse Compton losses are the dominant energy loss, then the higher energy electrons are selectively depleted, steepening the high frequency part of the spectrum. If this mechanism has progressed to affect the entire radio spectrum available from the ground, then a spectral break will occur at a lower energy, producing a flatter emission spectrum below 10 MHz; i.e. a simple extrapolation of the observed power law will not be valid.

3.0 RAMIFICATIONS OF EXTRAPOLATING TO $\gamma = 100$

Suppose that none of the effects discussed in section 2 are important, and that we may safely extrapolate the "observed" power laws to lower energies.

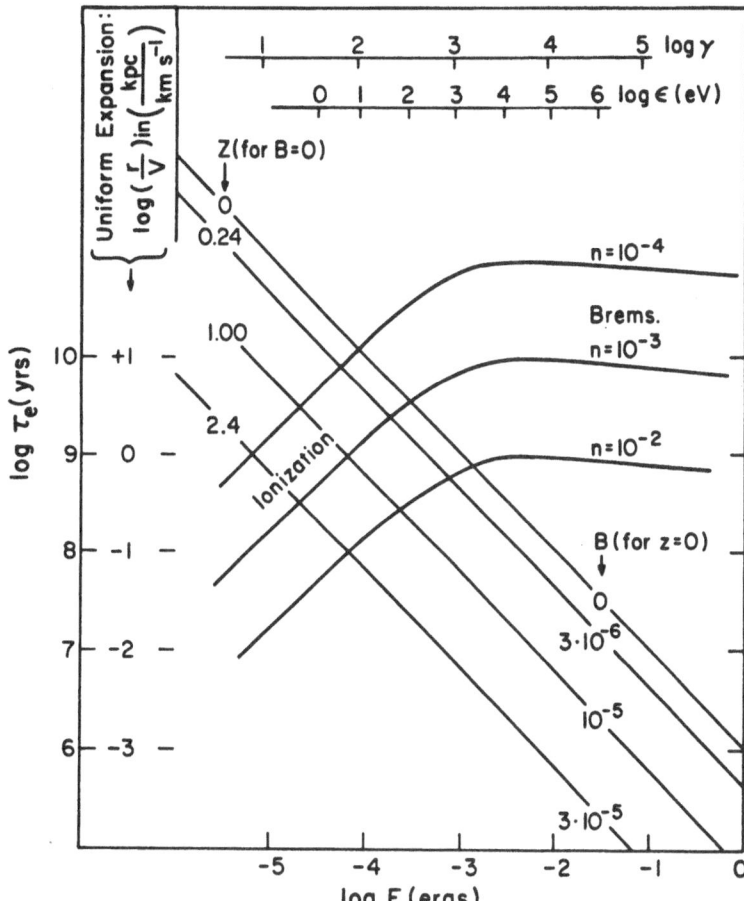

Figure 3. Half-life of Electrons (at the source). The ionization and brems-strahlung are combined into curves labelled with the ambient density. The lines of slope -1 are for E² losses, serving either for synchrotron losses for z=0 (with B indicated) or for IC/3K losses for B=0 (z indicated). The values for uniform expansion show how long it takes for the radius to double: the r value is the initial radius, and after the indicated time, all electrons will have lost half their energy. As in fig. 1, a scale for the IC/3K observed emission is added at the top (log ϵ).

3.1 Were Steep Spectrum Radio Galaxies Once Fanaroff-Riley class II Objects?

For this class of sources, the morphology is often that of large volumes and low surface brightness (lobes and bridges). These conditions are those for which weak magnetic fields strengths are found, which means that despite the relatively low luminosity observed, a large number of electrons are required to produce the radio emission. Harris et al. (1988) have compared the inferred number of electrons at low energies for Cygnus A to those for two low luminosity radio galaxies. For the range of interest here ($100 < \gamma < 1000$), the low luminosity galaxies contained as many electrons as Cygnus A. Since these electrons have very long lifetimes in low field regions, they can be used as an indicator for the original "power" of the source.

3.2 Constraints on the Acceleration Mechanism

The statistical processes currently under study as astrophysical models require that particles be "injected" with relativistic energies. Eilek and Hughes (1990), in their review of particle acceleration, discuss this problem. They, and references cited therein, use injection energies of γ from 100 to 800. If this is the case, then we may expect that the electron energy spectrum will have a low energy cutoff in that range (c.f. the model for the hotspots in Cygnus A by Carilli et al. 1990). Determination of the emission spectrum below 10 MHz should allow us to differentiate between various absorption mechanisms in the emission spectrum, and a cutoff in the electron spectrum (less inverted than $\alpha = -0.3$). Determination of injection energies would provide us with a new parameter, critical to understanding the acceleration process.

4.0 OTHER METHODS OF SAMPLING THE ELECTRON SPECTRUM

Besides synchrotron radiation, there are two other loss channels which produce radiation. We briefly discuss how they could relate to the study of the electron distribution at low energies.

4.1 Inverse Compton - 3K Emission

The relativistic electrons responsible for radio synchrotron emission will also scatter 3K background photons, boosting their frequency by γ^2. Since the maximum energy density of background photons occurs near $\nu = 1.6 \times 10^{11}[1 + z]$, the observed intensity at a particular observing frequency on Earth will arise chiefly from a particular energy electron, regardless of the redshift of the source. Thus, electrons with gammas between 100 and 1000 will produce IC/3K emission between 10^{15} and 10^{17} Hz, a spectral band in which observations of high sensitivity and resolution are not easy to make. Because of the available technology, searches for IC/3K emission have concentrated on the soft x-ray band, and have been unsuccessful, mainly because of the seemingly omnipresent thermal emission from hot diffuse gas. As resolution and sensitivity at higher energies improve, the thermal competition diminishes and the prospects become much better.

Dewdney et al. (1990) report on a radio halo in an uncatalogued cluster of galaxies at z=0.13. This halo has a low surface brightness and is of order 1 Mpc in size, leading to an equipartition magnetic field strength of $0.6\mu G$. From Einstein observations with the imaging proportional counter, an upper limit to the x-ray flux is found to be 4×10^{-13} erg cm^{-2} s^{-1} (0.5-3 keV). If the actual field is roughly the equipartition value, then there must be enough electrons to produce the observed synchrotron emission so that they would also produce $f_x = 6 \times 10^{-14}$, a value that can be reached with future x-ray missions. If this effort is successful, x-ray observations from 0.2-2 keV, will sample the electron spectrum between $\gamma = 580$ and 1700.

4.2 Bremsstrahlung

For this type of loss, the electron loses most of its energy in a single collision, and the emission band thus corresponds to the electron energy band: for electrons with γ's of 100 to 1000, γ-rays will be produced in the 50 - 500 MeV band. Although the EGRET experiment on the Gamma Ray Observatory covers this range, strong radio sources such as the lobes of Cyg A will probably not be detectable, even if the ambient density in the lobes = 10^{-3} cm^{-3} (Harris, 1989).

5.0 SUMMARY

There are both theoretical and observational indications that the electron spectra inferred from radio observations can not be extrapolated to γ values much below 1000 without a break or cutoff in the power laws found at higher energies. The capability of making reliable observations in the decade below 30 MHz would allow us to evaluate loss mechanisms other than the standard E^2 losses, and would also permit tests of acceleration models.

Although the main theme of this meeting was the science which could be done by space observations, it seems reasonable to conclude with a few remarks on instrumentation. As always, we require high resolution and high sensitivity. Because of the horrendous interference environment in near Earth orbit (and it can only get worse), it seems to us that the backside of the moon is the place to be. Because of the difficulty of delivery and operation at that site, we believe that a simple "6C" type of interferometer is the most promising concept: one element in orbit around the moon, observing so long as the Earth is below the horizon, with one or more elements on "luna firma". In the meantime, while we plan for the eventual lunar observatory, we should build the best possible Earth-based observatory in the best possible location to cover the 5 to 50 MHz band. Together, these two installations would allow us to obtain the scientific results discussed at this meeting.

References

Braude, S. Ya., Megn, A.V., Sokolov, K.P., and Sharykin, N.K. 1979, *Astrophys and Space Sci.*, **64**, 127.

Carilli, C.L., Perley, R.A., Leahy, J.P., and Dreher, J.W. 1990, *Ap.J.* (submitted).

Dewdney, P.E., Costain, C.H., McHardy, I., and Willis, A.G., Harris, D.E., and Stern, C.P. 1990, *Ap.J.Suppl.* (to be submitted)

Eilek, J.A. and Hughes, P.E. 1990, in Astrophysical Jets, Cambridge University Press, P.E. Hughes, ed.

Harris, D.E. 1987, in Radio Astronomy from Space; Proceedings of NRAO Workshop #18, K. Weiler, ed. p. 265.

Harris, D.E., Dewdney, P.E., Costain, C.H., McHardy, I., and Willis, A.G. 1988, *Ap.J.*, **325**, 610.

Harris, D. E. 1989, in Proceedings of the Gamma Ray Observatory Science Workshop, Goddard Space Flight Center, Greenbelt MD; W. Neil Johnson, ed. p 4-131.

Pacholczyk, A.G. 1970, Radio Astrophysics, W.H. Freeman & Co., San Francisco.

RADIO EMISSION FROM INTERGALACTIC GAS, AND ITS IMPLICATIONS FOR LOW FREQUENCY ASTRONOMY IN SPACE

Philipp P. Kronberg
Department of Astronomy, University of Toronto

I. INTRODUCTION

Synthesis mapping of extragalactic systems at frequencies below 1.4 GHz has revealed important new clues about what we might discover at much lower frequencies in intergalactic space. State-of-the-art images at frequencies between 1420 MHz and 300 MHz can be considered important precursors to even lower frequency images such as might be generated by space– or ground-based antenna arrays. In this talk I shall discuss how recent images at 327 and 408 MHz provide us with new information and some interesting clues about the physical phenomena which could be explored if we had sensitive, even lower frequency images at comparable or better resolution.

The recent outfitting of both the NRAO Very Large Array and the Westerbork Synthesis Telescope (WSRT) with 327 MHz receivers has permitted us to generate images which combine the virtues of high dynamic range, good resolution and high sensitivity at a relatively low frequency. The DRAO Synthesis Telescope at 408 MHz, and the Cambridge 151 MHz array (also described at this meeting) though less sensitive, have wide field capability which is also important for some types of low frequency imaging which we would like to attempt at lower frequencies.

In the following, I shall describe results of detailed mapping of intergalactic radio emission at 1.4GHz, 408MHz, and 326 MHz in the region of the Coma cluster of galaxies. The results provide some new information on intergalactic magnetic fields, and reveal large scale synchrotron emission at 327 MHz on *intercluster* scales for the first time. They point to potentially interesting discoveries that could be made in future at lower frequencies, where the diffuse intergalactic synchrotron emission will be even stronger.

The Coma cluster, due both to its richness and relative proximity, is the most accessible "passive laboratory" for studying the intracluster medium of a cluster of galaxies. However, even for the Coma cluster the properties of its halo have been known in less than adequate detail. This is partly because most radio telescopes tuned to relatively low radio frequencies (at which radio halos are most prominent) have inadequate resolution, poorly known beam characteristics, or are insensitive to the faint, large scale radio emission.

II. THE OBSERVATIONS AND RESULTS

a) Combined-array 1.4 GHz Observations of the Coma Cluster

To combine the advantages of good resolution and sensitivity to large scale structure, K.T. Kim, myself, P.E. Dewdney and T.L. Landecker carried out a series of new observations using the NRAO Very Large Array (VLA), and the Dominion Radio Astrophysical Observatory (DRAO) interferometer at Penticton, Canada, to produce a high dynamic range radio image of the Coma cluster which has good resolution *and* sensitivity to the extended structure. These observations have been described in some detail by Kim *et al.* (1990), and can be briefly summarized as follows:

The Coma cluster was mapped with the DRAO Interferometer at 1420 MHz, and at 1380 MHz with the NRAO VLA. The two datasets were combined and, using image optimization techniques, it was possible to produce a detailed map of the Coma halo with a dynamic range of 3000:1 down to a sensitivity level of 30 μJy/beam with a resolution of approximately 1 arcminute. At this resolution it is possible to distinguish most individual radio sources from the radio halo, and thus to isolate the true *intra-cluster* component of the radio emission. Figure 1 shows the 1.4 GHz radio map. Note that the resolution is sufficiently high that individual discrete radio sources are seen distinctly from the more diffuse cluster halo radiation.

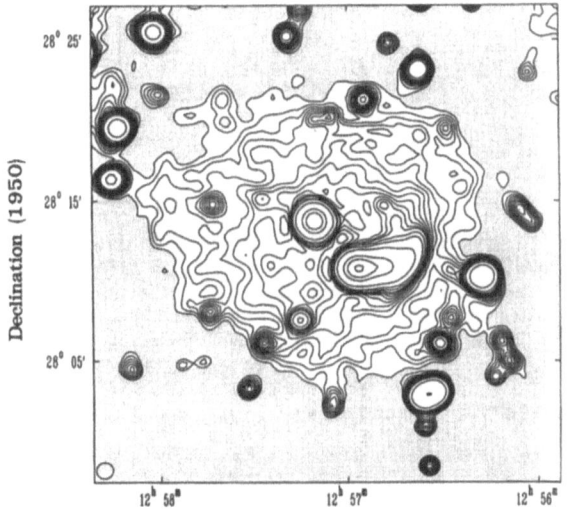

Right Ascension (1950)

FIGURE 1. The contour map of the radio halo in the Coma cluster of galaxies observed at 1.4 GHz. Contours are shown at −0.6, 1.2, 1.4, 1.6,...., 3.4, 4, 5, 10, 50, 100, 200 mJy/beam. The restoring beam is 71″×60″ elongated at −85.5° (reproduced from Kim *et al.* , 1990).

In another series of observations with the VLA, we conducted polarimetric mapping at four frequencies of 18 discrete radio sources at varying angular distances from the centre of the Coma cluster. The frequencies were 1465, 1665, 4835 and 4885 MHz, and were suitably spaced so that the Faraday rotation of the integrated polarized emission, (if the emission was polarized) could be determined with minimal ambiguity. It proved possible to determine rotation measures (RM) for 17 sources, and a further 18th was obtained by combining polarization data obtained elsewhere. Figure 2 illustrates that there is an excess of Faraday rotation measure at the position of the cluster halo. Kim *et al.* (1990) have combined this data with an EINSTEIN X-ray image by Abramopoulos and Ku (1983), in addition to further radio polarization data to infer the tangling scale of the magnetic field. The combined results were used to estimate that the magnetic field strength in the Coma cluster is of order 2 μgauss.

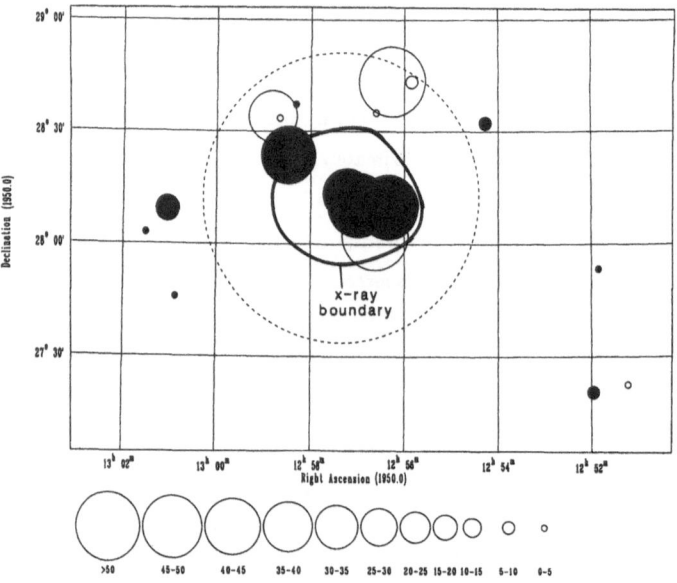

FIGURE 2. The RM distribution is shown for 18 sources seen through the Coma cluster and its vicinity. Filled and unfilled circles represent positive and negative RMs, respectively, and the sizes of circles are proportional to their absolute RM values, as shown in the bottom of the Figure. The dashed circle is centered on the cluster and has a radius of 39 minutes of arc, the Abell radius of the Coma cluster. The solid line is the approximate boundary of the EINSTEIN IPC X-ray source (see text). (reproduced from Kim *et al.* , 1990)

b) Wide-field DRAO-Effelsberg 408 MHz maps of the Region Around the Coma Cluster

A 4° by 4° area centered on the Coma cluster was mapped with the DRAO interferometer at 408 MHz, giving an angular resolution of 7.4′ by 3.4′. The wide field of view is possible due to the small size (9 m.) of the DRAO interferometer antennas. To these data were added short baseline information (i.e. the lowest-order Fourier spatial components) from the Effelsberg 100 m. telescope by Haslam *et al.* (1982). The resulting map, reproduced from Kim *et al.* (1990) is shown in Fig. 3. It is sensitive to structure on all scales at a sensitivity limit of 8 mJy/beam. When compared with the 1.4 GHz map, it also permits us to investigate variations in spectral index over the cluster halo. Kim and Giovannini will shortly publish what is the most detailed spectral index of a galaxy cluster halo, after combining these results with data at 1.4 GHz (above) and 327 MHz data at similar resolution (see below). A preliminary spectral index map (Figure 4), obtained from comparison of the present 408 MHz and 1420 MHz maps (Kim, 1988) illustrates pronounced spectral steepening in some parts of the Coma halo −− which are due to a combination of "aging" and diffusion of the relativistic electrons responsible for the emission. This result has tantalizing implications for low frequency astronomy of the future, in that it suggests that images of comparable quality at even lower frequencies will probably trace "old", steep energy spectrum relativistic electron populations at greater distances out into extragalactic space. This surmisal is now substantiated with new WSRT data at 327 MHz, which will be described next.

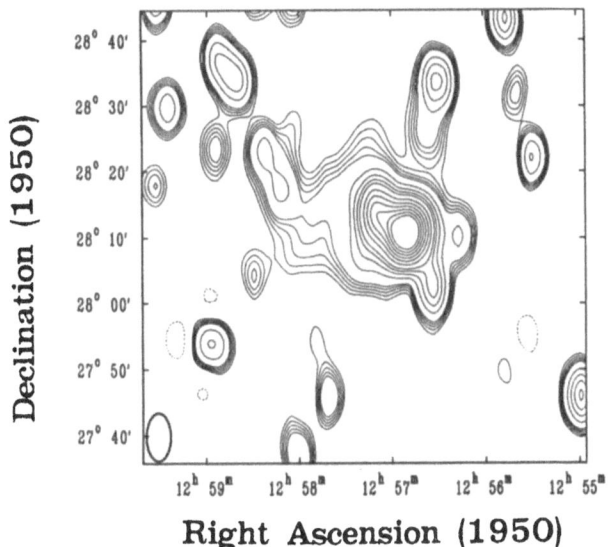

Right Ascension (1950)

FIGURE 3. Full synthesis radio map of the Coma C region at 408 MHz, made with the DRAO telescope, after adding data from the Effelsberg 100m. telescope. The synthesised beam size (shown in the lower left corner) is 7.45′×3.45′ and the sensitivity is 8 mJy/ beam. (reproduced from Kim *et al.* , 1990).

c) 327 MHz Synthesis Mapping of part of the Coma-Abell 1367 Supercluster

A portion of the Coma supercluster plane was observed for a total of seventy two hours in May and June 1986 at 326 MHz, using the Westerbork Synthesis Radio Telescope (WSRT); --see Venturi *et al.* (1989) for a description of the observing procedure. The observations were made in redundancy mode (Noordam and de Bruyn (1982)) to increase the dynamic range. The synthesized beam of the final map was 2′×1′, elongated N-S, and the rms noise level of the CLEANed map is 1.1 mJy/beam. The useful field of view covers an area of 2.6° in radius around the field center.

Additional VLA observations at 327 MHz in the A and B configurations, and centered both on the cluster center and on 1253+275, were used to find and subtract discrete sources present in the "bridge" region between the two galaxy clusters. The resulting map, convolved to a lower resolution (4′NS×2′EW), is shown in Figure 4. It shows a diffuse "bridge" of faint emission at 327 MHz which extends in projection beyond the Coma cluster halo in the direction of Abell 1367 for a distance of ∼1.5h_{75}^{-1} Mpc in projection. The residuals, after subtracting out discrete sources, are sufficiently well-determined that we can rule out the possibility that the "bridge" region is an artifact of the low-order interferometer spacings. Furthermore, with a 1 − 2 arcminute resolution which, at the distance of the

Coma cluster, is comparable to a galaxy size, we can rule out the possibility that the diffuse intergalactic emission is simply a blend of discrete radio sources.

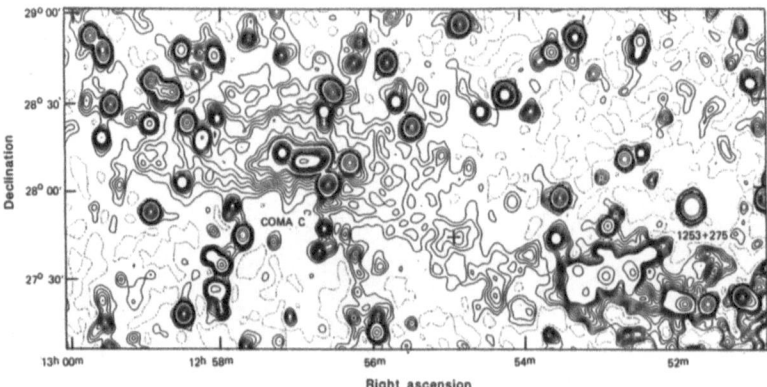

FIGURE 4. A field around the Coma cluster of galaxies was observed with the WSRT at 327 MHz by Kim *et al.* (1989). 10 antennas were fixed, and the 4 moveable ones were set at "SB" + n72m, where n=0 to 37, and "SB" (the shortest baselines) were 35, 48, 60, 72, 84, and 96 for six separate 12-hour sessions. To enhance the signal-to-noise ratio of the diffuse emission, small diameter sources, which are seen on the "bridge" region, were all removed and the residual map was convolved with a beam of resolution of 4'×2'. The projected linear size of this source is about $1.5h_{75}^{-1}$ Mpc. The location of NGC 4839 is marked with a + symbol. The peak surface brightness is 9.1 mJy/beam. Contours are shown at -1,1,2, ... 9,10,20,30, ...90,100,200,400 times 4 mJy/beam. (reproduced from Kim *et al.* 1989).

The spectrum of this bridge emission, though not known precisely, has been determined to be steep (Kim *et al.* , 1989) $\alpha \sim -1.5$ $(S \sim \nu^{\alpha})$, which confirms that it must be synchrotron radiation, and hence is a tracer of a large scale intergalactic magnetic field. The total flux density, S, contained in the bridge region is 760±100 mJy at 326 MHz. Again, we expect that much more of such emission would be detected at an order of magnitude lower frequency, say 30 MHz.

III. DISCUSSION

The foregoing maps at 1420, 408 and 327 MHz all reveal diffuse intergalactic emission which, having a generally steep spectrum, will be all the more detectable at very low radio frequencies. The 327 MHz maps show that magnetic fields appear to pervade the i.g. space in the westward extension of the Coma cluster halo. To the extent that widespread synchrotron emission can be detected at low radio frequencies, it has an associated equipartition magnetic field, B_{eq}, given by

$$B_{eq} = 0.21 \left(\frac{S}{0.75Jy}\right)^{2/7} (1+k)^{-2/7}\phi^{-2/7} \left(\frac{d}{1Mpc}\right)^{-2/7} \text{ microgauss.} \quad (1)$$

Since we do not know if the "bridge" in Fig. 4. is more like a tube, or an edge-on sheet in 3-space, we shall leave the line-of-sight dimension of the radiating volume as an unknown, d. ϕ is the volume filling factor, and k the (unknown) ratio of the proton to electron energies. S is the flux density received from within the angular boundaries in question. From Equation (1), taken from standard synchrotron radiation theory we see that the inferred equipartition magnetic field strength is relatively insensitive to the unknown quantities k and d. The quality of the image may not be sufficient to deduce ϕ accurately, however B_{eq} is not very sensitive to ϕ. Since source blending due to inadequate resolution is likely to give if anything an overestimate of ϕ, a detection of diffuse synchrotron radiation will tend to underestimate B_{eq}. Thus, we can conclude that the widespread magnetic field associated with the extended "bridge" in Fig. 4 has a strength (assuming equipartition) of a fraction of a microgauss.

Since S increases rapidly at low frequencies for steep spectrum emission, it is obvious that high dynamic range, and sensitive radio maps with enough resolution to separate compact from extended emission is a powerful, and perhaps the best, method of searching for weak intergalactic magnetic fields.

Of course low frequency radio astronomy detects only those magnetised regions which are "illuminated" by cosmic ray electrons to produce synchrotron radiation. The other side of this coin is that such observations, particularly at the lowest feasible frequencies, are sensitive tracers of old, or "fossil" CR electrons.

Another interesting aspect of low frequency intergalactic "glow" detections is that the associated field strengths, being sub-microgauss, suggest that inverse Compton losses will probably overwhelm synchrotron losses. Where CR densities are sufficiently high, inverse Compton-generated X-rays will result by scattering of the electrons off microwave background photons. Here we see the potential for an important complementarity in the future between l.f. radio and X-ray techniques for probing the physics of intergalactic gas. This is particularly interesting given the prospect of the ROSAT and AXAF satellites, which will greatly improve our current ability to detect faint diffuse intergalactic X-ray emission.

Apart from that recently demonstrated evidence by Kim et al. (1989) for intergalactic 327 MHz "glow", there is another physical reason for attempting comparable maps and surveys at even lower frequencies: It results from the fact that the lifetimes of electrons radiating near 327 MHz are not much greater than a few times 10^8 yrs. If these i.g. electrons were originally accelerated in galaxies, and diffuse into i.g. space with velocities typical of a few hundred km/s, then the *distance* they travel in this time is not much greater than the dimensions of a galaxy. That is, they are not the oldest of fossils, and they will not be the *most widespread* tracers of cosmic ray gas (and magnetic fields). Of course, we should keep in mind caveats to the diffusion velocity argument above, in that there may be other, effective ways of producing energetic electrons *in situ* at greater distances from the origin of the cosmic rays, such as pair production resulting from primary protons, which can travel much greater distances without much energy loss (Dennison, 1980).

It would also be interesting to probe the distribution of cosmic ray electrons at least back to 10^9 years. The characteristic losstimes of cosmic ray electrons is, scaled to 327 MHz, given approximately by

$$\tau \sim 3.10^8 \left(\frac{\nu}{327\text{MHz}}\right)^{-1/2} \text{years.} \tag{2}$$

This illustrates that, if a map comparable to that in Fig. 4 were made at 10 MHz, we would be able to trace 'fossil" CR electrons in weak intergalactic magnetic fields which were produced over 10^9 years ago, near the galaxy formation epoch.

At frequencies around or below 10 MHz, we may in practice have difficulty with absorption by the galactic interstellar plasma. A high dynamic range map as in Figure 4, but at 10 MHz would probably be impossible to produce from an earth based telescope because of *ionospheric effects*. There are two attractive possibilities, which I would not regard as mutually exclusive: In fact we should envisage a two stage "game plan". First, a compromise frequency between 30 and 75 MHz for an *earth-bound* array, such as a suitably enhanced VLA, or the new Indian radio telescope at Pune. The second would be to go to the lowest possible frequency which avoids excessive interstellar plasma absorption (perhaps in the range 6 - 10 MHz), and establish it on the *far side of the moon* in order to be free of terrestrial radio interference.

Financial support is acknowledged from the Natural Sciences and Engineering Council of Canada (NSERC).

REFERENCES

Abramopoulos, F., and Ku, W. H.-M. 1983, *Ap. J.,* **271** 446.

Dennison, B. 1980, *Ap. J.,* **239.** L93.

Haslam, C. G. T., Salter, C. J., Stoffel, H., and Wilson, W. E. 1982, *Astron. Astrophys. Suppl.* **47,** 1.

Kim, K.-T. 1988, *Ph. D. Thesis,* University of Toronto.

Kim, K.-T., Kronberg, P. P., Giovannini, G. and Venturi, T. 1989, *Nature* **341,** 720-723 .

Kim, K.-T., Kronberg, P. P., Dewdney, P. E., and Landecker, T. L. *Ap. J.* (in press) (May 1990).

Noordam, J. E., and de Bruyn, A.G. 1982, *Nature* **299,** 597-600.

Venturi, T., Feretti, L., and Giovannini, G. 1989, *Astron. Astrophys.* **213,** 49-60.

IN AND AROUND EXTRAGALACTIC SOURCES AT LOW FREQUENCIES[1]

T.W. Jones
University of Minnesota, Department of Astronomy
116 Church Street, S.E.
Minneapolis, Minnesota 55455

Introduction

I was asked to talk about radio jets in the context of the Low Frequency Space Array. A few nearby sources (e.g., Cen A) or very large sources (e.g., NGC6251) may actually have jets that can be resolved in at least one dimension on scales $\sim 10"$ (see the review by Bridle and Perley 1987, for example). However, most extragalactic sources would not reveal much about their structures on those scales, even ignoring the effects of interstellar and interplanetary scattering. So for the most part, we must be content to look for interesting physics that can be derived from the integrated properties of the sources.

To see what might be possible it is useful to start by establishing some guidelines. To be specific about what can be observed, I will use characteristics of the space array discussed by Weiler et al (1988). Then we can think in terms of sources that have fluxes $> \sim 1$ Jy around 30 MHz, and somewhat stronger at lower frequencies. Assuming a 10" resolution, we are thus limited to sources that have brightness temperatures in excess of $\sim 10^7$ K. By comparison, the galactic synchrotron background has a brightness temperature at these frequencies $\sim 10^5$ K. Thermal sources that might in principle satisfy this observability constraint would include hot plasma in a cluster (possibly heated by a passing high speed jet), and perhaps similarly hot gas in the nuclear regions of active galaxies. Much more likely to be observed, of course, are the relativistic electrons seen through synchrotron radiation. At these low

---

[1]An invited review presented at the Low Frequency Astrophysics from Space workshop, sponsored by NRL at Crystal City, Virginia, Jan. 8-9, 1990.

frequencies we ought to be alert, as well, to the potential for coherent emission processes of various kinds. Gas colder than $\sim 10^7$ K, though not observable in emission, could be detected through absorption, as it is in our own galaxy (Namir Kassim and Ron Reynolds, for example, have discussed that phenomenon at this workshop. See also, Kassim 1989).

Free-free absorption by "cold" gas in and around a radio source may begin to play a role in some cases, as we shall see. At high galactic latitudes, we probably do not have to be too concerned about effects of gas in the Milky Way. If there is sufficient nonrelativistic gas mixed in the radio emitting regions, dielectric suppression of the synchrotron emission (the Razin-Tsytovich effect) may also alter the observed spectral form of sources.

Since we are talking about wavelengths $>\sim 10$ meters, most intrinsically polarized sources will be strongly affected by Faraday rotation. Only sources exhibiting very small rotation measures, $\ll 1$ radian m^{-2}, stand a reasonable chance of not being completely depolarized and, thus, being suitable for polarization studies. Since typical rotation measures are >10 radian m^{-2} (Simard-Normandin and Kronberg 1974), I will not discuss that subject further.

Plasma Processes at Low Frequencies

To evaluate various possible plasma effects we need to determine some canonical source physical parameters. Using standard formulae (e.g., Pacholczyk 1970; Jones, O'Dell and Stein 1974), for example, we can evaluate the equipartition magnetic fields to be

$$B_{eq} \approx 5 \times 10^{-6} \ (1+k)^{2/7} \left[\frac{F_\nu}{Jy}\right]^{2/7} \left[\frac{arcsec}{\theta}\right]^{4/7} \left[\frac{30 \ kpc}{R}\right]^{2/7} \quad \text{Gauss.}$$

I have assumed here that the synchrotron spectral index is $\alpha \approx 0.75$ and that the low frequency emission cutoff is $\nu \approx 10$ MHz, although the value of B_{eq} is not very sensitive to those choices. Likewise, we can estimate the frequency at which these sources become synchrotron self-absorbed to be

$$\nu_1 \approx 1 \ (1+k)^{1/21} \left[\frac{F_\nu}{Jy}\right]^{8/21} \left[\frac{arcsec}{\theta}\right]^{16/21} \left[\frac{30 \ kpc}{R}\right]^{1/21} \ MHz.$$

Here, I assumed that ν_1, where $\tau_{ssa} = 1$, is the same as the low frequency cutoff, and that $B = B_{eq}$. The quantity k is the ratio of energy in invisible matter to visible electrons. I have expressed as many quantities as possible in terms of observables. Other than k, only the line of sight dimension of the source, R, is beyond direct measurement. R = 30 kpc corresponds to roughly 1" projected on the sky, if the source redshift $z \sim 1$ (for $H_0 = 75$ km s^{-1} Mpc^{-1}). For simplicity, I have ignored redshift corrections to each of these expressions, since they will not greatly alter the flavor of what follows.

From these expressions it is clear that typical sources that are nearly resolved will become opaque to synchrotron self-absorption in the vicinity of a few MHz, and that they have equipartition magnetic fields of several microgauss to several tens of microgauss. Under these circumstances, one is then observing radiation at low frequencies, $\nu \approx 10$ MHz $\approx \gamma^2 \nu_B$ corresponding to $\gamma \sim 10^3$, which is coincidentally roughly the same as one observes in compact variable sources at centimeter wavelengths. This means that the characteristic brightness temperature will be $\sim 10^{12} (\nu_1/\nu)^{11/4}$ K, assuming $\nu_1 < \nu$. For a 1 Jy source we can see that the effective size (radiating area) will be ~ 0.7" $\nu_1^{-91/64}$. This adds some theoretical insight to the comment earlier, that few sources will be resolved at low frequencies.

Spectra of radio galaxies sometimes exhibit increased slopes at high frequencies (e.g., Kellermann, Pauliny-Toth and Williams 1969). This presumably reflects either the effects of radiative energy losses for electrons emitting at high frequencies or alternatively results from increased efficacy of acceleration processes for lower energy electrons. Might we hope to learn something special about either of these from observations at a few MHz? As a crude model for acceleration let us take a generic Fermi process in which particles which diffuse with a characteristic diffusion coefficient, κ, through a medium with a characteristic velocity field, u. The velocity field might be turbulent, in which case the process is second order, or it might be representative of shocks, in which case the process is first

order. Either way the characteristic time to accelerate particles
will be

$$t_{acc} \sim \kappa/u^2.$$

We will want to compare that time to a characteristic dynamical time
for the radio source, $t_{dyn} \sim R/u$, which for $R \sim 100$ kpc and $u \sim 0.01 -$
0.1 c will fall somewhere around 10^7 yrs or more. To be specific
let's take the scattering to be from resonant interaction with Alfven
waves, and apply the usual quasilinear theory (e.g., Bell 1978). Then

$$\kappa \approx c \, r_1 \quad ,$$

where $r_1 = p \, c/(e \, B)$ is the Larmour radius for a particle of momentum
p, and $\delta B/B$ is the fractional field amplitude of Alfven waves with
$\lambda \approx r_1$. This leads to an acceleration timescale

$$t_{acc} \sim 200 \left[\frac{c}{u}\right]^2 B_{\mu G}^{-3/2} \left[\frac{B}{\delta B}\right]^2 \text{ sec.}$$

Adopting equipartition field values, t_{acc} will be very short compared
to t_{dyn} for particles of momenta $p \sim 10^3$ mc, such as those observed at
low frequencies. However, in the same environments, it is just
conceivable that the acceleration times for electrons with $p \sim 3 \times 10^4$
mc which would radiate near $\nu \sim 10$ GHz, would require $t_{acc} \sim t_{dyn}$ if the
MHD turbulence is relatively weak; namely, $\delta B/B <^\sim 10^{-3}$. The
radiative timescales under the same circumstances are

$$t_{rad} \approx \frac{10^{10}}{\nu_7^{1/2} B_{\mu G}^{3/2}} \text{ yrs ,}$$

where ν_7 x 10 MHz is the critical frequency of electrons emitting
synchrotron radiation in a magnetic field $B_{\mu G}$ microGauss. Once again,
the expected lifetimes for particles radiating at low frequencies do
not match up very well with either acceleration or dynamical times,
whereas, there is an interesting similarity in all three if $\nu_7 \sim 10^3$
(10 GHz). Consequently, as has been noted many times before (e.g.,

Bridle and Perley 1987) observations above and below a few GHz are important to our understanding of the relativistic electron histories in large radio sources. It is not so clear that low frequencies add much to that, however.

Earlier, I touched on the possibility of synchrotron self-absorption in extended radio sources. Although free-free absorption through our own galaxy seems unlikely, at least well away from the plane, there may be some possibility that free-free absorption in and around the sources themselves is observable. One instance might derive from the emission line clouds that are now being found around some radio lobes (McCarthy and van Breugel 1989). These may be related to the matter that seems to strongly Faraday depolarize some sources, but that is not yet clear. From the emission lines we cannot tell if the clouds are in front, inside, or behind the radio lobes. Perhaps low frequency observations can help resolve that issue. As a simple model, suppose the emission line gas is in clouds of density n and radius r, occupying a volume of size R with a filling factor $f = N\, r^3/R^3$, where N is the number of clouds. The associated covering factor is $\omega = f\, R/r$. Then the free-free optical depth through a single cloud is

$$\tau_{ff} \sim 4\ L_{\beta 41}\ T_4^{-3/2}\ \nu_7^{-2}\ R^{-2}\ \text{kpc}\ \omega^{-1},$$

expressed in terms of the total luminosity in the H_β line, $L_{\beta 41} \times 10^{41}$ erg s^{-1}. That is related to the local parameters using $L_\beta \sim 10^{-25}\ n_e\ n_H\ V\ T_4^{-1}$ erg s^{-1} [e.g., Osterbrock (1989)]. For comparison, the emission lines detected so far fall into the luminosity range $\geq 10^{40}\text{-}10^{44}$ erg s^{-1}, so it would seem possible that these clouds might obscure the radio sources, at least in part. If $\tau_{ff} \gg 1$, then the clouds will block a fraction ω of the radio emission from behind if the clouds are in front of the source and a fraction $\sim\omega/2 + f$ if the clouds are randomly mixed within the radio source. However, unless this fraction is ~ 1, so that a near exponential emission cutoff occurs for intervening clouds, it would seem to be difficult to distinguish between the two.

On the other hand, for this thermal gas there is another effect that may show up if the gas is actually mixed with the relativistic plasma; namely, dielectric suppression, or the Razin-Tsytovich effect. That results from the fact that in a plasma an electromagnetic wave with $\nu > \nu_p$ has a phase speed greater than the speed of light. This means that in terms of their radiant properties fast particles don't appear to be moving so relativistically. Accordingly, the power they radiate is reduced whenever $\gamma > \nu/\nu_p$, or $\nu < \nu_{RT} = 20\ n_e/B_{\mu G}$ MHz. At low frequencies there is an exponential cutoff (e.g., Pacholczyk 1970). Once again, however, if the plasma density to magnetic field intensity ratio is high only in some regions, the net effect may be less dramatic. We can directly compare the frequencies at which the Razin-Tsytovich effect and free-free absorption become important for our emission line clouds. In particular,

$$\frac{\nu_{ff}(\tau = 1)}{\nu_{RT}} \sim 1 \quad \frac{T_4^{-3/4}}{B_{\mu G}} \frac{f}{\omega} R_{kpc}^{1/2}$$

Clearly, this leaves us with the possibility that either free-absorption or Razin-Tsytovich is more important in a given situation. Note that this does not depend upon the plasma density, but only on the temperature and magnetic field as local parameters.

One of the effects that is an intriguing possibility at low frequencies is, of course, coherence. This is more likely at low frequencies for several reasons. In the case of stimulated emission, this is because the ratio of stimulated to spontaneous emission coefficients varies as λ^3. In the case of coherence due to phase bunching in the orbits of radiating particles it is because the number of particles that can act coherently generally scales also as λ^3. Many mechanisms have been suggested over the years to result in coherence, including synchrotron radiation (e.g., Zheleznyakov and Suvorov 1968), "ripple radiation" (which is related) (Cocke, Pacholczyk and Hopf 1978), and a slew of plasma processes (see Kaplan and Tsytovich 1973, for some examples). Let me outline one interesting possibility in the latter class. Consider the hot gaseous region that presumably confines emission line clouds associated with the "narrow-line clouds" (NLC) in QSOs. The NLC themselves have characteristic densities, $n_e \sim 10^{6-7}$ cm^{-3} and temperatures, $T \sim 10^4$ K. Pressure balance requires the confining medium, therefore, to satisfy $n_e T \sim 10^{10-11}$. If we take the hot, intercloud medium to have temperatures $T \sim 10^8$ K (e.g., Krolik, McKee and Tarter 1981), the electron density $n_e \sim 10^3$ cm^{-3}. Thus the electron plasma frequency is $\nu_p \approx 3 \times 10^5$ Hz. If the hot plasma is turbulent, as one might reasonably expect in such an environment, a substantial amount of plasma energy may be channeled into Langmuir waves and low harmonics (which are electromagnetic modes). The strength of this turbulence can be expressed as a parameter β, describing the ratio of nonthermal Langmuir wave energy to thermal Langmuir wave energy (see, e.g., Colgate, Lee and Rosenbluth 1970, Jones and Kellogg 1972), or equivalently the ratio of electron density fluctuations due to the same two causes; namely, $\beta \approx (\Delta n^2)/(\Delta n^2_{th})$.

When a photon scatters off a Langmuir wave, momentum and energy conservation require $\mathbf{k}' = \mathbf{k} \pm \mathbf{k}_p$ and $\nu' = \nu \pm \nu_p$, respectively. Thus we have a kind of photon Fermi process. In a thermal plasma the scattering cross section for this interaction is just the Compton cross section. In a turbulent plasma, however, the scattering will be enhanced by the presence of the Langmuir waves. In principle, as many as $\sim n k_D^{-3} \sim 3 \times 10^8 \, n^{-1/2} \, T_4^{3/2}$ electrons (the number in a Debye sphere) can scatter coherently. In the hot intercloud medium of an AGN narrow line region, that number would produce an enhancement in the scattering cross section over the Compton value, σ_T, by a factor $\sim 10^{11}$. Because of the constraints of the momentum conservation relation, the requirement that Langmuir waves have phase velocities greater than the electron thermal speed, which is less than the speed of light, the scattering tends to be preferentially in the forward direction, and preferentially to higher frequencies (Colgate, Lee and Rosenbluth 1970, Jones and Kellogg 1972). Consequently, the available solid angle is reduced to something like $\Omega \sim (k_D/k)^2$. The total cross section for scattering the photons is then

$$\sigma \sim \beta \, \sigma_T \, (k_D/k)^2,$$

with the Debye wavenumber, $k_D \approx 1.4 \times 10^{-3} \, n^{1/2} \, T_4^{-1/2}$ cm^{-1}. I have used β as the estimate for the coherence factor. In such an environment we would expect photons generated at low frequencies, just above the plasma frequency, to be upscattered roughly until the scattering depth $\tau \sim n \sigma R < 1$. Taking $R \sim 100$ pc, so that $n \sigma_T R \sim 0.2$, the maximum expected frequency for coherent Langmuir upscattering ought to be $\nu_7 \sim 0.3 \, \beta^{1/2}$. With a modest degree of turbulence, so that $\beta \geq 10^2$ this would fall in the range of observation of radio telescopes. Of course, it would have to be the case that the power put into this turbulence was sufficient to generate an observable flux. But the energy available is $\sim\beta \times 10^{56}$ erg, so that is not unreasonable. Because the gas is so hot, $T \sim 10^8$ K, free-free absorption will not prevent this coherently produced emission from escaping at frequencies greater than $\nu_7 \sim 1$.

Conclusions

There are, in fact, a large number of processes which can modify the spectra and appearance of radio sources at low frequencies. That is certainly why we have had this meeting -- to see if we could take advantage of these special conditions to learn something about the sources that we could not otherwise hope to do. The problem may be that there are so many processes that it will turn out to be difficult to sort out which are important in a given situation. Others at this meeting have dwelt upon influences from the interplanetary and interstellar media, so that is not my immediate concern. Rather, I need only to look to the sources themselves to face that prospect. We have seen that synchrotron self-absorption is likely to cause source spectra to turn down or at least flatten at a few MHz. But, free-free absorption from enveloping or incorporated clouds may do the same, as may the Razin-Tsytovich effect. Defining practical ways to discriminate among the possibilities when one observes a source turnover at $\nu \sim 10$ MHz will not be easy.

I think the primary information about electron acceleration and radiative losses is probably going to come from higher frequencies, especially if one expects any of the above spectrum-modifying processes to be active.

I have always been intrigued by the possibility of observing coherent emission from extragalactic objects at low frequencies. Pulsars have taught us that coherent radiation is at once very helpful in gaining some basic theoretical insights about source physics, but also theoretically very challenging. The most likely signature that we are observing coherence will be time variability. But, because of strong scintillation effects at low frequencies, that may be challenging as well.

Finally, I must add the rather trite, but nonetheless valid, comment that when extensive observations are made at these frequencies from space, the most interesting results will be the ones that aren't mentioned here or anywhere else, yet.

Travel support to attend this meeting came from the NSF through grant AST87-20285 and from the Minnesota Supercomputer Institute.

References

Bridle, A. H. and Perley, R. A., 1987, Ann. Rev. Astr. & Ap. (v 22), 319.

Bell, A. R., 1978, M. N. R. A. S., 182, 14.

Cocke, W. J., Pacholzyk, A. G., and Hopf, F. A., 1978, Ap. J., 226, 26.

Colgate, S. A., Lee, E. P., and Rosenbluth, M. N., 1970, Ap. J., 162, 649.

Jones, T. W., O'Dell, S. L., and Stein, W. A., 1974, Ap. J., 188, 353.

Kaplan, S. A., and Tsytovich, V. N., 1973, Plasma Astrophysics (Pergamon Press: Oxford).

Kassim, N. E., 1989, Ap. J., 347, 915.

Kellermann, K. I., Pauliny-Toth, T.I.K., and Williams, P.J.S., 1969, Ap. J., 157, 1.

Krolik, J. H., McKee, C. F., and Tarter, C. B., 1981, Ap. J., 249, 422.

McCarthey, P. J. and van Breugel, W., 1989, Proceedings of the ESO Workshop on Extranuclear Activity in Galaxies, Garching, (to be published).

Osterbrock, D. E., 1989, Astrophysics of Gaseous Nebulae and Active Galactic Nuclei (University Science Books: Mill Valley, CA), p. 74.

Pacholczyk, A. G., 1970, Radio Astrophysics (W. H. Freeman: San Francisco).

Simard-Normandin, M., and Kronberg, P. P., 1974, Astr. & Ap. Suppl., 43, 19.

Zheleznyakov, V. V., and Suvorov, E. V., 1968, Soviet Physics JETP, 27, 335.

Weiler, K. W., et al, 1988, Astr. & Ap., 195, 372.

LIST OF PARTICIPANTS

Baldwin, John E.	Cavendish Lab., Cambridge, U.K.
Basart, John P.	Iowa State U., Ames, IA, U.S.A.
Bridle, Alan H.	N.R.A.O., Charlottesville, VA, U.S.A.
Burns, Jack O.	New Mexico State U., Las Cruces, NM, U.S.A.
Carr, Thomas D.	U. Florida, Gainesville, FL, U.S.A.
Chevalier, Roger A.	U. Virginia, Charlottesville, VA, U.S.A.
Cordes, James M.	Cornell U., Ithaca, NY, U.S.A.
Crane, Patrick C.	N.R.A.O., Socorro, NM, U.S.A.
Dennison, Brian K.	Virginia Tech., Blacksburg, VA, U.S.A.
Desch, Michael D.	NASA/GSFC, Greenbelt, MD, U.S.A.
Dewey, Rachel J.	JPL, Pasadena, CA, U.S.A.
Dickey, John M.	U. Minnesota, Minneapolis, MN, U.S.A.
Dulk, George A.	U. Colorado, Boulder, CO, U.S.A.
Erickson, William C.	U. Tasmania, Hobart, Tasmania, AUSTRALIA
Farrell, William	NASA/GSFC, Greenbelt, MD, U.S.A.
Fey, Alan L.	C.A.S.S., NRL, Washington, DC, U.S.A.
Fiedler, Ralph L.	C.A.S.S., NRL, Washington, DC, U.S.A.
Geldzahler, Barry J.	A.R.C., Landover, MD, U.S.A.
Gergely, Tomas E.	N.S.F., Washington, D.C., U.S.A.
Gopalswamy, N.	U. Maryland, College Park, MD, U.S.A.
Gupta, Yashwant	U.C.S.D., La Jolla, CA, U.S.A.
Harris, Daniel E.	C.F.A., Cambridge, MA, U.S.A.
Jackson, Bernard V.	U.C.S.D., La Jolla, CA, U.S.A.
Johnston, Kenneth J.	C.A.S.S., NRL, Washington, DC, U.S.A.
Jones, Dayton L.	JPL, Pasadena, CA, U.S.A.
Jones, Thomas W.	U. Minnesota, Minneapolis, MN, U.S.A.
Kaiser, Michael L.	NASA/GSFC, Greenbelt, MD, U.S.A.
Kassim, Namir E.	C.A.S.S., NRL, Washington, DC, U.S.A.
Kronberg, Philipp P.	U. Toronto, ON, CANADA
Kuiper, Thomas B.H.	JPL, Pasadena, CA, U.S.A.
Kundt, Wolfgang	I.f.A., Bonn U., F.R.G.
Lebo, George	U. Florida, Gainesville, FL, U.S.A.
Lind, Kevin R.	C.A.S.S., NRL, Washington, DC, U.S.A.
Longair, Malcolm S.	Royal Obser., Edinburgh, Scotland, U.K.
Magnani, Loris A.	Arecibo Obser., Arecibo, PR, U.S.A.

LIST OF PARTICIPANTS (continued)

Odegard, Nils P.	NASA/GSFC, Greenbelt, MD, U.S.A.
Onello, Joseph	SUNY, Cortland, NY, U.S.A.
Payne, Harry E.	STScI, Baltimore, MD, U.S.A.
Phillips, J.A.	Arecibo Obser., Arecibo, PR, U.S.A.
Radhakrishnan, V.	Raman Res. Inst., Bangalore, INDIA
Reber, Grote	Bothwell, Tasmania, AUSTRALIA
Reiner, Michael	NASA/GSFC, Greenbelt, MD, U.S.A.
Reynolds, Ronald J.	U. Wisconsin, Madison, WI, U.S.A.
Safaei-Nili, Ali	Iowa State U., Ames, IA, U.S.A.
Siqueira, Paul	Iowa State U., Ames, IA, U.S.A.
Smith, Eric P.	NASA/GSFC, Greenbelt, MD, U.S.A.
Smith, Harlan J.	U. Texas, Austin, TX, U.S.A.
Spangler, Steven R.	U. Iowa, Iowa City, IA, U.S.A.
Spencer, John H.	C.A.S.S., NRL, Washington, DC, U.S.A.
Stone, Robert G.	NASA/GSFC, Greenbelt, MD, U.S.A.
Webber, William R.	New Mexico State U., Las Cruces, NM, U.S.A.
Wehrle, Ann E.	OVRO, Caltech, Pasadena, CA, U.S.A.
Weiler, Kurt W.	C.A.S.S., NRL, Washington, DC, U.S.A.